NATURE
CONSERVANCY
COUNCIL

Nature conservation and afforestation in Britain

Nature conservation and afforestation

Nature conservation and afforestation

Contents

Introduction

Forest is the natural climax vegetation over a large part of the World's land surface and supplies people with resources for many of their needs. The practice of forestry which manages and sustains these renewable resources is an ancient and honourable skill. Yet in many parts of the World our relationship with forests is out of balance. Man has destroyed forests on a vast scale, and continuing inroads into the remaining tracts of natural woodland especially in the humid Tropics and Subtropics cause world-wide anxiety. Britain itself is a region which, more than most, has cleared the original forests. Its agricultural prosperity, in particular, came about through the farming of land once covered in trees, though the use of wood for fuel and other purposes also contributed to the widespread destruction of forest.

By the end of the 19th century, Britain retained barely 5% of its surface under woodland, and it remains today, after the Netherlands and the Republic of Ireland, the least wooded country in Europe. In parallel with the removal of native forest has grown the depend-ence on overseas supplies of essential timber, both softwood and hardwood. Island isolation and the resulting precariousness of wartime timber supplies gave a great impetus to the state programme for reafforestation, which has forged ahead in this century with energy and success. The develop-ment of nuclear weapons has made less compelling the argument for a strategic defence reserve of timber but the domestic timber-producing industry has continued to expand with unabated vigour. The restoration of at least some of the lost woodland cover has been considered by successive governments as a prudent land-use policy represent-ing sound investment for the future, and both the state and the non-state forestry sectors have contributed to this expansion though the balance between their respective contributions has varied.

In parallel with those of farming, the methods of modern forestry have become increasingly sophisticated efficient and dependent on introduced species. The previous sense of need to safeguard food supplies as well as to increase those of timber placed one large and basic restriction on reafforestation — that it should not compete for good quality farmland. This restriction tended to conflict with the further requirement to maximise economic return, but this could in part be achieved by planting trees which gave the greatest timber production in the shortest time. Such factors determined that the afforestation programme should be concentrated on infertile soils, mainly in the hillier districts of Britain, using exotic softwood conifers. Management of existing lowland woods also became much influenced by the superior economic advantages of replacing slow-growing native hardwoods with conifers, especially introduced species.

Within the new plantations, foresters have shown increasing concern to plan and manage in the interests of landscape, amenity, wildlife conservation and recreation. However the process of reafforestation is not always in harmony with other, and rapidly growing, interests which value the countryside for its scenic beauty and landscape and for the enjoyment and study of wildlife. The Nature Conservancy Council has observed the developing situation of nature conservation and forestry for many years, but recently with increasing anxiety about some aspects of this relationship. Its views were expressed briefly in 1980 in evidence to the Sherfield Committee on *Scientific aspects of forestry* (House of Lords 1980). The Committee's Report has led to an improvement in conservation prospects for existing broadleaved woodlands through the Forestry Commission's recently introduced policy for broadleaves, and the Nature Conservancy Council looks forward with some confidence to the realisation of the new opportunities which this gives for furthering the conservation of ancient semi-natural woodland. Other initiatives, such as the Timber Growers UK *Forestry and woodland code,* are also welcomed.

Over new afforestation, however,

we continue to be very concerned. Several recent cases of conflict between forestry and nature conservation interests on important wildlife areas and the portents for continuing losses to wildlife and physical features have convinced NCC that it has now to make its views widely known, so that nature conservation needs in relation to afforestation are understood and recognised. Much has been said in the last ten years or so about the case for further afforestation, but there is no adequate statement about the relationship between afforestation and nature conservation. This paper aims to provide such a statement.

A debate is already under way on the future of Britain's countryside, now that most agricultural products are in surplus, and in consequence large areas of land may in the future no longer be required for producing food. There are many competing uses for this land, among them growing timber and nature conservation. Our concern is to see the right balance struck.

An early draft of this document was widely circulated and this revision has benefited from the comments we received. We are deeply grateful to all those organisations and individuals who have contributed. We believe that forestry and timber production are important for this country; equally we believe that because circumstances have changed, new policies are required which reconcile the economic and social objectives with those of nature conservation and amenity as harmoniously as possible. This document should be seen as our contribution to formulating these fresh policies and the surrounding debate. We intend our role to be a constructive one.

William Wilkinson

William Wilkinson
Chairman NCC
14 April 1986

The problem for nature conservation

Woodland in Britain varies enormously in nature conservation value (Figure 1). On the top of this scale is ancient semi-natural woodland of native trees (broadleaves and Scots pine), and at the bottom are mono-culture plantations of introduced trees. The woodlands presently being created on a large scale are mainly of the latter type, and our review deals with the nature conservation problems which arise from such afforestation, involving planting on ground which has long been treeless or almost so. The fundamental issue is that most modern afforestation causes an ecological transformation, in which the non-living components (notably soils and water) of the open ground ecosystems are significantly altered and the wildlife community of plants and animals characteristic of open ground is largely replaced by one characteristic of forest. The remnants of the open ground wild-life communities which persist within or periodically return to the new forest ecosystem are a totally inadequate representation of these communities.

The Timber Growers United Kingdom (TGUK) have expressed the situation succinctly in *The forestry and woodland code* (1985), under Afforestation and New Planting. "The introduction or re-introduction of forestry onto bare land will produce immediate changes in the general ecology of the area. The exist-ing habitat will largely disappear, and some species with it. New habitats will develop and new species will be attracted." Holmes (1979) also said: "In establishing forests on bare upland it is hard to imagine a more dramatic change in the ecological conditions including the soil, the flora, the fauna, and not least the introduction of trees for the first time in many centuries." The Centre for Agricultural Strategy's (1980) study made similar comments.

The crucial point is that the new forests are **quite different** from the heaths, grasslands, peat bogs and sand dune communities which they replace. **However good they may become as wildlife habitats in their own right, these forests are not and never can be a compensatory substitute for such open**

Figure 1 Variations in nature conservation value of forest according to human intervention.

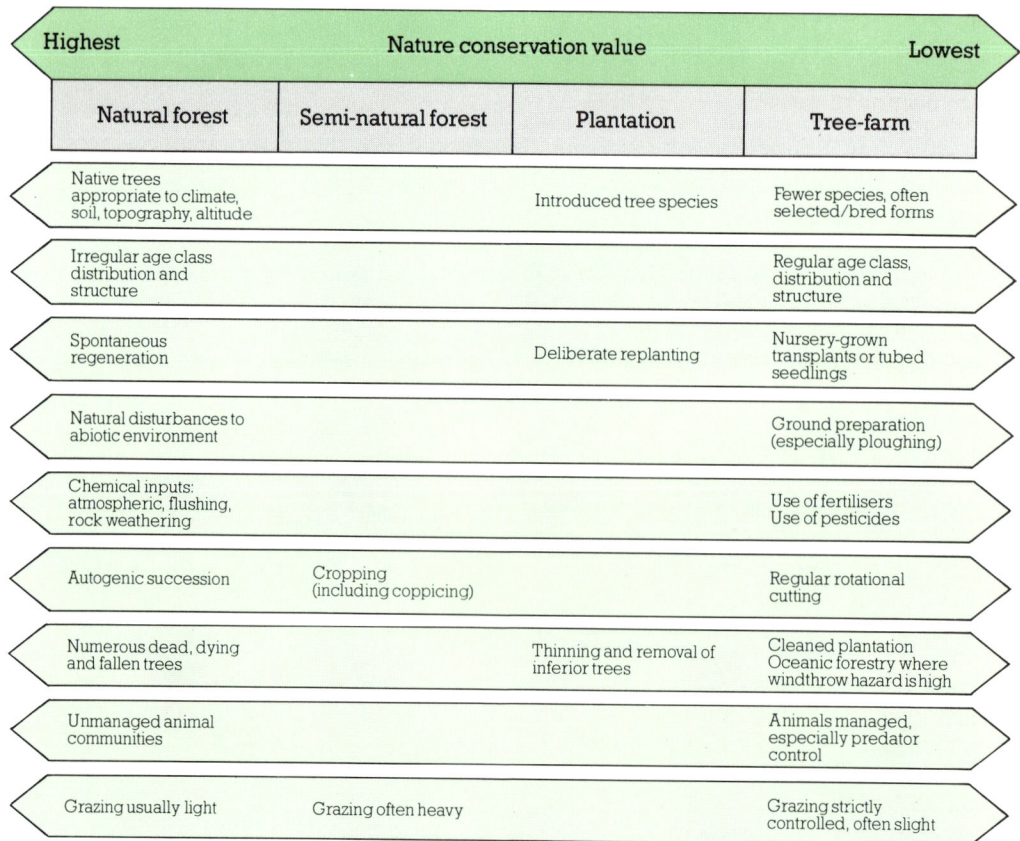

Highest	Nature conservation value		Lowest
Natural forest	Semi-natural forest	Plantation	Tree-farm
Native trees appropriate to climate, soil, topography, altitude		Introduced tree species	Fewer species, often selected/bred forms
Irregular age class distribution and structure			Regular age class, distribution and structure
Spontaneous regeneration		Deliberate replanting	Nursery-grown transplants or tubed seedlings
Natural disturbances to abiotic environment			Ground preparation (especially ploughing)
Chemical inputs: atmospheric, flushing, rock weathering			Use of fertilisers Use of pesticides
Autogenic succession	Cropping (including coppicing)		Regular rotational cutting
Numerous dead, dying and fallen trees		Thinning and removal of inferior trees	Cleaned plantation Oceanic forestry where windthrow hazard is high
Unmanaged animal communities			Animals managed, especially predator control
Grazing usually light	Grazing often heavy		Grazing strictly controlled, often slight

1

Nature
conservation
and afforestation

The problem
for nature
conservation

**ground ecosystems, which are highly
valued in their existing condition.**
Afforestation also tends to obliterate or
even destroy physical features of high
scientific interest. Nature conservation
therefore requires that a reasonable
and adequate balance be reached and
maintained between the extent and
distribution of the new forests and the
range of existing open ground types
and physical features outside the
forests. This review presents the
conservation case for such a balance. It
aims to state:

☐ the nature conservation value of open
  ground ecosystems;
☐ the impact of afforestation on their
  wildlife and physical features;
☐ the present problems of relationship
  between nature conservation and
  afforestation;
☐ the desirable measures for safeguarding an adequate total amount,
  distribution and variety of such
  ecosystems.

The achieving of such a balance
can come about only through discussion and agreement between all the
land-use interests concerned, including
not only forestry but other parties
closely involved with the rural
environment.

There has been little if any
consideration of the nature conservation value of the ground to be afforested,
and there has been a reluctance to
acknowledge that afforestation poses
any real problem for nature conservation. The wildlife gains have been
portrayed as a net benefit in the
ecological change, and the case has
been argued from the standpoint that
forestry concern for nature conservation begins when trees are planted or
forest already exists rather than when
the selection of a site for planting is
made. *The Forestry Commission and
conservation* (Forestry Commission
1980, rev. 1986) puts this view. New
forests have the potential to become
valuable wildlife habitats, but this has
been used to argue that the unselective
loss of open ground habitats through
afforestation is justified, since large
areas of such habitats still remain
unplanted in various regions of Britain.
Unplanted areas are also frequently

depicted as having low wildlife or other
value compared with the new forests.
As a result, a one-sided picture, lacking
the nature conservation viewpoint, has
developed in the public presentation of
the issues.

The present review has first to
redress this imbalance by explaining
the losses to wildlife and physical
features caused by afforestation. It does
not claim that all remaining open
ground is of equal importance in its
nature conservation value, or suggest
that there should be a halt to further
afforestation. Open ground ecosystems
vary widely in both the character and
the importance of their wildlife and
physical features, and the main
objective is to ensure that a sufficient
total area, including both the most
valuable examples and the overall
range of variation in types, remains
unplanted. The Nature Conservancy
Council (NCC) also have a statutory
duty, in the discharge of their functions,
to take account, as appropriate, of
actual or possible ecological changes
(Nature Conservancy Council Act 1973).
NCC must be concerned with the
general environmental impact of land-
use activities, and this includes interest
in the effects of afforestation on all
organisms and the physical-chemical
environment, especially soils and
water, within catchments containing
new forests.

The mountains and moorlands of
the north and west, where nearly all
new planting is taking place, are the last
substantial areas of wild land in Britain,
where both scenic beauty and wildlife
can be studied and enjoyed as an inter-
related whole in settings free from the
grosser intrusions of human activity. It is
true that much — though not all — of this
terrain was forest-clad once, and lost its
tree cover through clearance by man.
Much of the forest destruction was
completed before the Roman occupa-
tion, and even in Scotland most of the
loss occurred at least 400 years ago.
The derived vegetation is often dwarf
shrub heath which has changed little in
its botanical composition from the field
communities which still occupy the
forest floor of surviving remnants of the
ancient native woodlands. Such vegetation

7

The problem for nature conservation

forms much of the grouse moor habitat which holds one of the most varied and important upland animal (especially bird and insect) communities. There are also large areas of wet blanket bog which are naturally treeless, and some of these have intricate networks of pools and hummocks in a quite undisturbed condition. Many of these submontane vegetation types are unique to Britain, and the patterned bogs are regarded as especially important by ecologists. They support breeding bird communities representing the southernmost European occurrences of the Boreal-Arctic avifauna of Eurasia. The open ground habitats are also the haunt of a notable group of predatory birds, occurring here at unusually high density, compared with many parts of the world. Britain has, accordingly, an international obligation in the conservation of these open ground ecosystems.

A point often made by foresters is that they are only restoring forest cover to land which was once mostly under forest and has since been cleared and then substantially degraded by centuries of grazing, burning and other modification. In the case of the wetter peatlands, this is simply not so, and the grouse moors, deer forests and sheep-walks have acquired a considerable wildlife value in their own right, as a range of semi-natural ecosystems in which the flora and fauna are still largely of native species.The restoration of tree cover on ground of lowest wild-life value, with native trees and using methods not disturbing to the physical environment, would benefit nature conservation. But modern afforestation is quite different. It brings the techniques of arable agriculture to land which, for the most part, has never been under cultivation before; and it is appropriately described as 'tree farming' (Davies 1979). The ground is usually ploughed or ripped, fertiliser is often applied (and sometimes later renewed), and nursery-grown seedlings of exotic tree species are planted in straight lines. Herbicides and insecticides are not used on nearly the same scale as in agriculture, but they

Figure 2 The continuum between natural and artificial habitats.

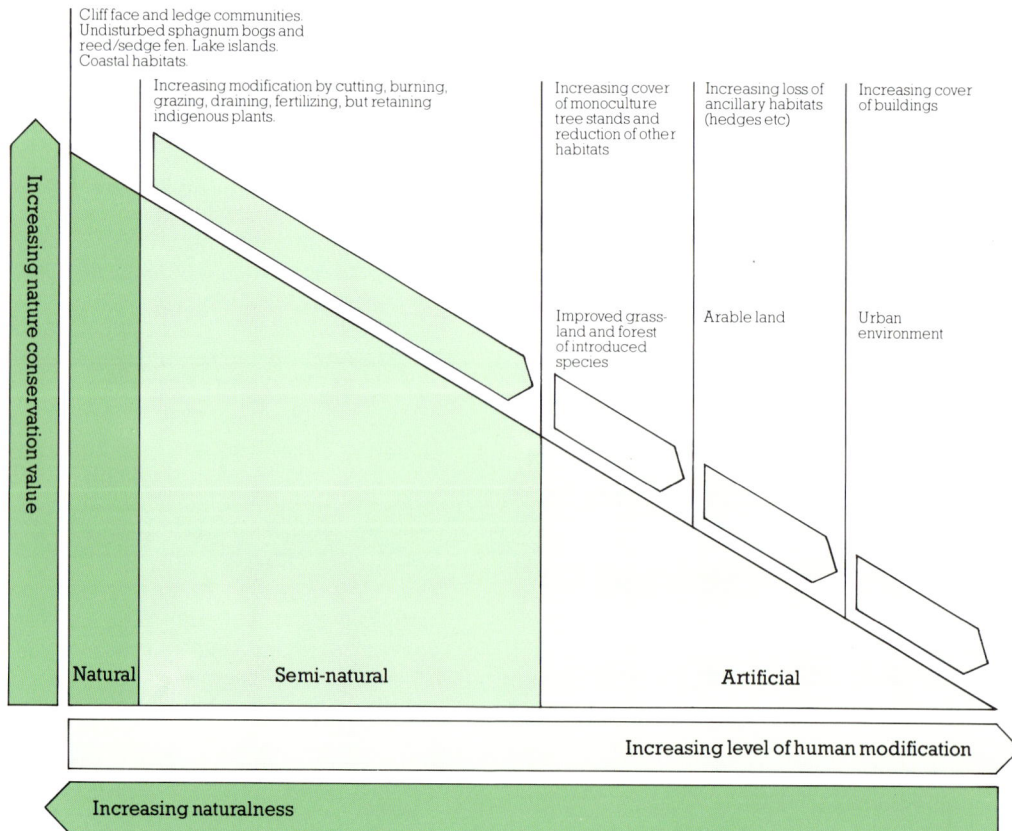

Cliff face and ledge communities. Undisturbed sphagnum bogs and reed/sedge fen. Lake islands. Coastal habitats.

Increasing modification by cutting, burning, grazing, draining, fertilizing, but retaining indigenous plants.

Increasing cover of monoculture tree stands and reduction of other habitats

Increasing loss of ancillary habitats (hedges etc)

Increasing cover of buildings

Improved grass-land and forest of introduced species

Arable land

Urban environment

Increasing nature conservation value

Natural

Semi-natural

Artificial

Increasing level of human modification

Increasing naturalness

8

are nevertheless often introduced to a previously almost pesticide-free environment. There are changes in the physical nature of the ground surface and the soil chemistry is altered. The pattern of water drainage and run-off also become quite markedly changed, and both sediment and nutrient loads typically become increased. Enhanced transfer of atmospheric acidity to waters occurs, and freshwater fish and other aquatic organisms may also be adversely affected.

Some new forests have good wild-life value, especially for animals, and there has recently been an encouraging concern to manage them with a view to future improvements, particularly in subsequent rotations. This is very welcome, and we would wish to help and support efforts to increase the value of the new forests as a wildlife resource. However, these forests have a lower place on the scale of naturalness than either the habitats they replace or ancient semi-natural woodlands composed of native trees (see Figure 2). This situation is likely to continue because to attempt the re-creation of forest more truly akin to the natural types of the Boreal regions on any significant scale would be economically and logistically impractical.

NCC believes that, within the remaining area of Britain that could be planted with trees, there is scope for satisfying the needs of both nature conservation and forestry. The maximum planting rates contemplated by the forestry interest would still leave a large extent of open ground in the mountains and moorlands, whilst the new possibilities for extending afforestation to the more fertile, lower ground of the agricultural regions would seem to give even more chance of reducing competition between nature conservation and forestry. And the more that such lowland planting could be with native broadleaves, the more it would be welcome to nature conservationists. Such a sharing of the land resource will, however, require much goodwill and determination to find practical solutions to the problems inherent in the present arrangements under which the two sides

independently pursue their equally legitimate objectives.

9

Nature conservation policy and practice

In 1984 the Nature Conservancy Council published *Nature conservation in Great Britain* (NCGB), a review of progress and a statement of strategy for the future, taking account of the role and objectives of both the official bodies and the non-governmental organisations for nature conservation (NGOs). That document should be consulted for an exposition of overall objectives and methods. In the present review there is an account only of nature conservation policy and practice regarded as particularly relevant to the afforestation issues to be examined. It is given largely from the perspective of NCC but tries to make appropriate reference to the position of the NGOs.

Methods of nature conservation

There are two basic approaches to the conservation of nature in Britain, whether by the official or the non-official bodies. The first is the special protection of the most important areas for wildlife and physical features by some designation which allows these areas to be managed primarily for this interest, that is as nature reserves. The second is the application of more general measures to the wider environment outside these special areas, including laws on species protection, but consisting mainly of advice and persuasion to all those whose activities impinge on this wider scene, in others words everyone, but especially landowners and occupiers. Every other activity, in public relations, education, information, research, funding and administration, supports the two basic approaches. Site of Special Scientific Interest (SSSI) is a designation applied to important areas by NCC which, though having legal force, falls short of conferring nature reserve status, in that existing land-use continues as the primary concern but is subject to a measure of influence designed to safeguard the nature interest (Department of the Environ-ment 1982; NCC 1983). SSSIs accordingly belong to the first approach rather than the second. Nature reserves include statutory reserves, namely the National Nature Reserves (NNRs) established or agreed by NCC and Local Nature Reserves established by local authorities, and non-statutory nature reserves established by other conservation bodies — notably the Royal Society for the Protection of Birds (RSPB), the Woodland Trust and the Trusts for Nature Conservation — and by individuals. The two National Trusts also have extensive land holdings which are managed for their scenic and wildlife value. SSSI designation now includes all NNRs and can be extended to all other nature reserves of the requisite standard.

The rationale for the specially protected area as the cornerstone of nature conservation strategy received a classic exposition in the seminal *Conservation of nature in England and Wales,* Cmd 7122 (Wild Life Conservation Special Committee 1947). This stressed that the reserves be viewed as a national series containing an adequate representation of the total countrywide range of ecological variation in major habitat types, with their representative flora, fauna and non-living environmental features — the series as a whole to include the characteristic and typical as well as the rare and unusual. A parallel representation of geological and physiographic features for their own scientific importance was also recommended. The purpose of this national reserve series was conceived especially as scientific study, research and education, but also as promoting the more general enjoyment, inspiration and awareness to be derived 'from the peaceful contemplation of nature'.

The NNR acquisition programme followed these precepts, but after a time it became clear that the lists of recommended reserves in Cmd 7122 were based on patchy and incomplete survey, and that, if consistent standards were to be applied, a great many more areas qualified for selection, especially in mountain and moorland districts. An attempt to locate further sites of NNR quality for their biological features led to the production of *A nature conservation review* (NCR) (Ratcliffe ed. 1977). This contained a description of 735 sites in this category, covering 913,400 hectares (4% of Britain) and including nearly all the existing NNRs.

During the period since the original Nature Conservancy was set up in 1949,

Nature conservation policy and practice

at least two important changes have occurred. The most significant is the enormous increase in the number of people who are well informed and have a caring concern for nature. This has caused the various parts of the spectrum of interest in nature and its conservation to become very differently weighted. The 'scientific' end of this spectrum, while still remaining important, is now substantially eclipsed in size by the 'popular' end. Government, as well as NCC, has become responsive to this shift in public concern. The other shift has been the increasing recognition of the United Kingdom's obligation to international conservation, especially by giving special attention to the safeguarding of those features which are exceptional and diminishing on the global or continental scale, and to migratory species. This commitment has been strengthened through several formal international agreements, for example the Ramsar Convention on conservation of wetlands, the Bern Convention on conservation of European wildlife and natural habitats, the Bonn Convention on migratory species, the Washington Convention on trade in endangered species of wild fauna and flora, and the EEC Directive on the conservation of birds, especially listed species and their habitats and also migrants.

A third and negative change is that the whole range of land use activities, i.e. agriculture, forestry, urban-industrial land and energy developments, water and mineral use, defence and recreation, have caused massive further attrition of the natural environment. The evidence on post-1945 loss and damage to the remaining national resource of wildlife habitats and physical features, together with the evident weight of public concern, persuaded Parliament to strengthen the legal measures for resisting further depletion in the Wildlife and Countryside Act 1981. The new SSSI provisions gave much more meaning to this designation, but also an added obligation for NCC to ensure that all areas of appropriate quality were identified and notified. The number of known sites of SSSI quality

for both biological and geological features is about 6,000, covering nearly 8% of Britain. NCC aims to notify all of these and regards some 614 of them as being of NNR quality in addition to the existing 214 NNRs (covering over 155,000 hectares). The NGOs have established nature reserves, and these now cover around 100,000 hectares, many of which are also SSSIs. Surveys, including assessment of hitherto neglected groups of plants and animals, continue and will result in the identification of additional important sites. NCC's estimate of the eventual total of protected areas using present selection criteria is under 10% of the total land surface, but future public interest will clearly have an influence on the eventual figure. Only a part of the total area of protected sites relates to afforestation (see p.73).

In the wider environment outside the protected areas of all categories, the conservation objective is to minimise further losses to the national resource of nature, which will continue through the competing claims of other users, including developmental interests (NCC 1984, section 15.2), and where possible to enhance the conservation interest. This involves seeking to ensure that in any one district a reasonable balance will be achieved between other land-use interests and nature conservation. The NGOs also aim to maintain wildlife habitats in a geographically even way, so that they remain available for people to study and enjoy countrywide. Afforestation is, again, only one of the land-uses concerned in this wider environmental programme.

The Countryside Act 1968 requires NCC to have due regard to the needs of agriculture and forestry and to the economic and social interests of rural areas, and nature conservation has to operate within a framework of existing land-uses. While often regarded as a land-use concern in its own right, wildlife interest is often closely related to traditional activities, and conservation problems tend to be the result of modern changes in use, often involving intensification, of which commercial afforestation is especially important.

Nature conservation policy and practice

Principles of site evaluation and selection

Afforestation has been concerned largely with uncultivated land of low agricultural value used mainly as rough grazing. Such land includes lowland grasslands and heaths and sand-dunes, but consists mainly of the great areas of open hill ground of grasslands, heaths and peat bogs in the north and west of Britain (Map 1). In the lowlands these habitats are mostly fragmented as isolated islands, but in the uplands they form large continuous areas, representing our last extensive expanse of wild country. Such open ground, with the scattered remnants of associated native woodland, has long been a particular focus of nature conservation concern, both in the protection of specific areas and in promoting more general measures.

It has become clear over the last 40 years that only within the specially protected areas is it likely that wild nature stands a reasonable chance of escaping further depletion. Except perhaps within the National Parks (and there are none in Scotland, where undeveloped upland is most extensive), nearly all other land and water must be regarded as highly vulnerable to further development. It is therefore vital that a minimum area of nature reserves or other designated land should be secured for posterity now. This requires that the series of protected sites representing the countrywide ecological gradients in wildlife habitats, and the range of physical features, be adequate in both number and extent. For each formation, a countrywide network of exemplary sites must be chosen to represent the overall geographical-climatic gradients and the main variations within these according to local geology, topography, soils and previous management. These sites should each meet certain minimum standards of quality, but the process of selection now adopts the principle that all sites above a specified standard should qualify for protection.

The NCR gave an account of the rationale for the selection of key biological areas, including basic principles and the criteria for assess-ment of candidate sites. The process of SSSI selection involves a further elaboration and refinement of the same approach and criteria, which NCC is currently revising and will publish in a more comprehensive statement. RSPB's nature reserves are chosen especially for ornithological value, those of the Woodland Trust for their woodland importance, and those established by local Trusts tend to be chosen for regional qualities. Threat and opportunity are important considerations in assessing priorities for reserve selection, but sites have to meet qualifying standards of intrinsic importance for wildlife.

The selection of exemplary or 'type' sites has to be related to reference points in a classification of the range of variation in wildlife and its habitats in Great Britain. Such a reference framework was given in *A nature conservation review,* and a comparable system for geological and physio-graphic features will be provided by the Geological Conservation Review currently in progress. The biological criteria for nature conservation evaluation are applied both to the definition of qualifying standards and to the comparative assessment of similar sites in choosing the best example of a particular type. The different site attributes mostly have additive value, so that the total 'score' is a measure of overall importance. It is, however, crucial to appreciate that only similar types of sites can be thus compared: an oakwood cannot be compared with a saltmarsh — only with other oakwoods. Moreover, criteria tend to shade into each other or actually overlap, rather than show clear-cut separation. The major criteria are naturalness, diversity, size, rarity, fragility and typicalness, the first four being especially important to the definition of minimum standards.

In Britain **naturalness** is a relative term, but naturalness is a continuum and nature conservation is especially concerned with the upper part of the range (G. F. Peterken, unpublished). Natural habitats are those not significantly modified by man. Tansley (1939), aware that a great deal of British vegetation was modified by human

Nature conservation policy and practice

activity, coined the term semi-natural to describe plant communities owing their character to some degree of such intervention, but remaining composed of native species and having structural features corresponding to those of natural types. The concept has since been extended to include modifications to the abiotic environmental conditions of soil, hydrology and microclimate. Figure 2 depicts the continuum of naturalness diagrammatically. Most of the open ground ecosystems with which afforestation is concerned span a range of variation within the semi-natural category, depending on the degree of modification by forest clearance, grazing, burning or other disturbance. But some of the peatlands — raised, valley and blanket bogs — are not deforested; they are naturally treeless because of the extreme wetness of their surfaces. Pollen analytical evidence indicates that the area occupied by the Caithness flow country never carried extensive woodland cover during the whole postglacial period (Peglar 1979). Some of these bogs are as close to the natural condition as any vegetation in this country, apart from that of cliffs, and they are highly valued accordingly.

Many of the grouse moors, with their prevailing dwarf shrub heaths (and often blanket bogs also), have communities of ling and bell heather, cross-leaved heath, bilberry, cowberry, crowberry and (in Scotland) bearberry which are the little-modified ground vegetation of earlier oak, birch and pine woods: they belong to the upper part of the semi-natural range. The same is also true of the lowland heaths so characteristic of acidic soils in southern Britain. The acidic grasslands derived by long continued grazing and burning from the dwarf shrub heaths, and variously dominated by purple moor-grass, mat-grass, heath rush, deer grass, fescues and bents, are at the lower end of the semi-natural range. Burning also favours the invasion of dry heath by bracken, but this fern (which is both an open ground and a woodland species) spreads still more readily within the derived grasslands. Though more modified than the dwarf shrub

heaths, these are distinctive semi-natural communities requiring representation. Although stable dune areas are subject to grazing, and heathy types may be burned as well, sand-dune systems are relatively natural. Communities composed largely of non-indigenous species are not semi-natural but artificial, and these do not qualify for selection. When two similar sites are compared, that with the lesser evidence of artificial features is preferable, other factors being equal.

Diversity is an important but often misunderstood criterion. It is a measure of richness or variety, of both communities and species, determined especially by underlying environmental variation; but it must be considered in relation to naturalness. The number of different habitat types within an area reflects diversity and is important. But diversity resulting from the presence, or creation, of artificial habitats adds to nature conservation value only when the other existing habitats are not of special interest: its effect is the reverse in areas highly rated for their existing features. In particular, the claim that exotic conifer planting helps to diversify the habitat is a distortion of the criterion, because such addition reduces naturalness. The creation of artificial diversity using non-native species and disruptive methods is essentially damaging to sites valued for their natural and semi-natural features (see also p. 51).

The assessment of finer-scale diversity, within a single habitat, is based on the variety characteristic of the habitat in question, some habitats being intrinsically more diverse than others. Sites qualify when they exceed the basic minimum for the type, and their importance increases as variety rises above this level, for example as measured by number of characteristic communities and species. In Britain, vegetation on acidic substrata tends to be species-poor and botanical variety usually increases with soil fertility, measured especially as rising pH and exchangeable calcium content. Because the open habitats favoured for afforestation are particularly on the poorer substrata which so predominate

Nature conservation policy and practice

in the British uplands, they tend to consist of widespread and species-poor vegetation, in which botanical variety depends greatly on physical habitat diversity, especially in hydrology. Even the poorest hills vary in drainage according to slope, giving a sequence from dry grassland or heath through moist types to blanket bog, as ground wetness increases. The more localised emergence of drainage water as springs, rills, flushes and flush bogs usually gives a marked increase in botanical variety, especially where the water has become nutrient-enriched by leaching more basic substrata below the surface (see Pearsall·1950). Localised patches of rich, drier grassland often result from intermittent flushing. Diversity in animal species (including invertebrates) is also valued. It often runs parallel to habitat and vegetational diversity, but not always.

Size is an important criterion, but especially in lowland districts, where semi-natural habitat now occurs mainly as numerous small island fragments, compared with the uplands, where such habitat forms large continuous expanses. In the former situation, the bigger the fragment, the greater its importance. Some habitats are now so reduced that virtually all remaining fragments deserve protection. In the uplands, concern is about adequacy of size, for the individual sites and for the series as a whole: the scale of occurrence has to guide the scale of representation. For upland sites, the essential need is to seek completeness of the features which each site is intended to represent, and this again involves considerations of diversity. Upland sites should comprise complete topographic-ecological units, not segments or slices. High mountains have the largest development of montane habitats and are important accordingly, but especially in relation to the whole range of more extensive submontane habitats forming the lower ground. In defining boundaries, it is thus important to include the complete mountain ecosystem, from the highest point down to the lowest levels of the adjoining valleys, and such areas often need to be large, in the range of 5,000 to 15,000 hectares.

The same arguments apply also to lower moorlands and peatland systems. For instance, in the flow country of Caithness and Sutherland it is not sufficient to select isolated patches of patterned bog: whole catchments need to be represented, with their drainage systems intact. Scientific needs are relevant here, for only by having large protected areas is it possible to have controls against which extensive and pervasive environmental changes elsewhere can be measured; for example acid deposition or hydrological disturbance.

The size of species populations is also relevant. Areas with large numbers of colonial animal species or high densities of non-colonial species rate highly, especially for those of restricted distribution. Protected areas should generally be large enough to ensure population viability for their component species, that is to say survival if all other populations outside these areas were lost. However, for the rarer or more highly dispersed species, especially large predators, protected areas may have to be considered in the aggregate, since few sites are individually large enough to hold more than a few pairs of each bird of prey.

One of the most important criteria is that of **rarity** — of habitats, communities and species. Rarity is generally also a measure of endangerment and hence of need for conservation priority. The rarer a feature, the greater is the risk of its disappearing unless specific protection is given. There is also a link with numerical status over time, in that a trend towards decline — as distinct from normal fluctuation — will tend to increase conservation concern, because unchecked decline is the prelude to extinction, either locally or nationally. This is another continuum, in which nature conservation concern increases along the gradient from common to rare. Rarity also links with naturalness, in that in Britain the most natural features are now quite uncommon and highly localised.

Fragility is a measure of the intrinsic sensitivity of natural features to human disturbance. Natural features

Nature conservation policy and practice

tend to be fragile, and the pervasiveness of human activity in Britain has caused the most fragile to become rare, so these criteria are inter-connected. Remoteness and inaccessibility have a mitigating influence, by reducing the element of threat and hence the resultant vulnerability of features. A good example is *Sphagnum*-covered bog surface, which is intrinsically fragile, being extremely sensitive to fire, lowering of the water table, trampling and nutrient enrichment. It is now a rare feature, steadily decreasing further and confined mainly to remote localities or protected areas.

Within the range of features represented in the series of protected sites, it is important that those which are **characteristic and typical** are included, as well as those which are unusual. This will normally be the case through application of the principle of representation; for example an area of high diversity will usually contain features typical of the district or region, and so will an upland massif chosen as

a topographical unit. It may, however, happen that such characteristic features are under-represented in protected sites, and that additional sites are required to fill the gap. Large areas of typical habitat are often necessary to sustain adequate populations of larger mammals and birds. Sometimes, uniformity of habitat (the opposite of diversity) can also be a desirable feature, especially for experimental research, which may need replication of plots under conditions of maximum available similarity.

The ecological attributes valued by these criteria are expressed in the important overall characteristic of **non-re-creatability**. This quality is, indeed, probably a better measure of nature conservation value than any other single factor or criterion. It could be regarded as an aspect of naturalness, and the more natural an ecosystem, the greater is the problem of re-creating it once it is lost. Restoring the physical conditions of former habitat is sometimes possible, but it is especially difficult to restore the complete and

Blanket bogs with pool and hummock systems formed of a living surface of *Sphagnum* moss are now very local in Britain. They are difficult to afforest but the drier parts can be drained and planted, and this leaves an inadequate remnant of the original bog system. Silver Flowe NNR, Galloway, 1956.

Nature conservation policy and practice

identical range, and even more problematical to replace the full species complement originally present. Many species of plant and animal are difficult to reintroduce, but the greater problem is that the full list of lowly species once present is simply not known. Re-created habitats thus tend to be regarded as second-best, at most, and great emphasis has to be placed on conservation of original ecosystems while and where th~y still remain.

The assessment of international importance involves application of the same criteria, and it especially concerns the occurrence of features which are rare on the global or continental scale. Britain is regarded as particularly important for features associated with an oceanic climate, of which ericaceous dwarf shrub heath and blanket bog are notable examples. Many of the vegetation types of the submontane zone, including those widespread kinds derived from earlier forests, have no counterparts anywhere else in the world. Heather moorland and heath have an increasingly frag-

mentary distribution within their limited climatic range in the west of mainland Europe, and lowland heaths of gorse and heathers in southern Britain have assumed increased importance since the destruction of those in Brittany. Blanket bog is a world rarity, confined to limited parts of a few hyper-oceanic regions, and Britain contains at least one-tenth of the total global area. The patterned surface systems well developed on many Scottish raised bogs and blanket bogs are of outstanding interest to peatland ecologists, and their breeding bird fauna represents a southern outlier of that on the tundras of northern Eurasia. Article 4 of the EEC Birds Directive requires Member States to implement special conservation measures for the habitats of certain listed resident species and regularly occurring migratory species. These include northern open ground species such as golden eagle, merlin, hen harrier, golden plover, green-shank, dunlin, common sandpiper and ring ouzel.

A patterned blanket bog surface. These systems of pools and intervening ridges and hummocks are especially characteristic of flat bogs ('flows') in northern and western Scotland. They are of great interest to those who study peatland develop-ment, and are inter-nationally important ecosystems.
Badanloch Bogs, Rimsdale, Sutherland, 1985.

Nature conservation policy and practice

Requirements for the national series of protected areas

The following brief account is limited to the kinds of habitat that are regarded as afforestable. To ensure adequate representation of habitats across the geographical-climatic range of Great Britain, the county — or district in Scotland — is taken as the basic area for site selection. Within each county where a major habitat occurs, sufficient examples should be chosen to represent the more local variations attributable to differences in geology, topography, soils, land-use or other factors. Where a county contains a large part of the total extent of a certain habitat or especially important areas, a larger than average county representation in protected sites is required. Representation of species assemblages or rare species may be covered by habitat selection but sometimes requires the choice of additional sites. Geological and physiographic features are chosen according to their own, separate criteria, but often overlap with sites of biological interest. Requirements for relevant habitats are summarised as follows.

Sand-dunes

These are widely distributed in coastal areas, so that a similarly dispersed series is desirable, covering the range of physiographic types. Large size is highly rated, and the range from acidic to calcareous should be represented regionally. Internationally important systems on the Hebridean machair and North Norfolk coast deserve special emphasis.

Lowland acidic heaths and grasslands

In the Breckland and on the Lizard serpentine, all remaining examples of the mixed grass-heath types should be protected, with no lower size limit. East of a line from Exmouth to Scarborough, all acidic heaths of more than 100 hectares should be included, along with any significant areas of associated wet heath and valley bog. West and north of this line, a geographically dispersed series should be represented, including southern Atlantic, maritime and sub-maritime, upland transition and Caledonian (north-east Scottish) types.

Peatlands

There are six main topographic classes of peatland, of which some valley mires, most raised bogs and all lower blanket bogs are relevant to afforestation. Valley mires are mainly a lowland type, and rather small in area, but they vary widely from acidic types to rich fen. A widely dispersed series of examples is needed, covering the range of edaphic variation in each district. Raised bogs are a highly localised, mainly lowland western type and are now much depleted in total area: all remaining examples larger than 100 hectares should be included (except when severely damaged), as should any smaller examples showing unusual features, such as those of structure. Blanket bog occurs widely from near sea level to well over 1,000 metres but is a particularly important feature in Britain because of its world rarity (see map 4). Raised and blanket bogs are both important structurally for the scientific interest of their peat deposits as a postglacial record and also for the small-scale micro-topography of their living surfaces. Patterned surfaces with networks of pools and hummocks are especially a feature of these peatlands in Scotland, and their variability according to situation and climate requires a widely dispersed geographical series for adequate representation. In east Sutherland and Caithness some 2500km^2 of low moorland is covered mainly with deep blanket bog — the 'flow country'. Any original tree growth was extinguished here long ago, largely by the natural development of the bogs under the cool, humid climate. The flat flows show an especially extensive and varied development of patterned surfaces, and there are numerous larger lochans and lochs. There is a breeding bird fauna especially rich in wetland and northern European species. This is a peatland and moorland ecosystem and landscape now unique in western Europe, and it represents the nearest equivalent here to the vast tundras of the Arctic regions. It has outstanding international importance (Ratcliffe ed. 1977; RSPB 1985; Prestt 1985), and special emphasis upon conservation measures is required there.

Nature conservation policy and practice

Upland grasslands and heaths

Blanket bogs are associated with these habitats in many mountain and moorland areas, but the present category refers mainly to drier ground. There is an extremely wide range of types, but these are classifiable into:

☐ sheepwalks, which tend to be grassland-dominated and occur on a wide variety of soils, but prevail on most of the more base-rich areas such as the Southern Uplands and Breadalbane Hills;

☐ grouse moors, which are usually heather-dominated and tend to lie mainly in the drier east; and

☐ deer forests,which are usually mixed grassland and dwarf shrub heath, especially in high and rugged mountain ranges of the Highlands and Islands.

Topography and soil fertility vary widely, and the botanically rich areas are mainly on the highly localised areas of calcareous rocks. Many higher mountains are valued for the extent of their montane habitats, above the potential tree line, where prolonged snow cover combines with extreme wind exposure and frost effects to give outliers of Arctic-Alpine fell-field and mountain tundra. Continuity with the lower zone of submontane habitats on these hills is important in showing the total altitudinal range and zonation, and the necessity of treating these massifs as topographic units has been stressed. Many of the widespread types of hill vegetation and associated animal communities in Scotland, and especially the submontane types, have evolved under a unique combination of climate and land-use. They are found only in Britain, though some belong to the class of Atlantic dwarf shrub heath, which is an especially localised and declining habitat in other parts of Europe (see p.16). British upland ecosystems cover an extremely wide range of variation and constitute by far the largest total area of natural or semi-natural types (about 7 million hectares). This needs to be reflected in the number and total extent of protected areas.

Combination of ancient and semi-natural woodland of oak and beech, lowland acidic heath and valley mire. One of the few extensive areas of semi-natural habitat remaining in southern England.
Denny Wood, New Forest, Hampshire, 1975.

Afforestation policy and practice

The creation of the Forestry Commission (FC) in 1919 was a response to the realisation that Britain, as a largely deforested country, was critically short of a strategic reserve of timber. The intention was two-fold: that the Commission would have a role as the State Forestry Authority, in promoting overall forestry policy and helping landowners to contribute to expansion of the timber resource, and as the Forestry Enterprise, in itself managing existing woodland and creating new forests. State forestry policy sought to avoid competition with agriculture, notably by concentrating afforestation on low quality land used only as rough grazing, if at all. Large tracts of such land were available, enabling forests to be established at low unit costs. FC and the Agriculture Departments developed a working relationship over the availability of plantable land, and the Commission has been able to buy large areas of poor land with low potential for agricultural improvement at relatively low prices. Land has been purchased for afforestation as it has become available on the open market, usually through the sale of hill farms, deer forests and grouse moors.

State expenditure on afforestation was complemented by encouraging private investment, which included special grant-aid and taxation incentives. These concessions encouraged afforestation by land-owners and, in more recent years, have led to the creation and growth of several private forestry companies, acting mainly as agents for others wishing to invest money in forestry by buying and selling suitable land and planting and managing forests. The system operates mainly through land purchase and forest establishment by high income earners; these benefit especially from tax incentives and can then convert their investment into capital by selling to the secondary market of corporate investors, who acquire an almost tax-free long-term asset (Moore 1985a). FC dispenses not only grant-aid but advice and practical help to the non-state sector. The afforestation methods of both sectors are essentially the same.

The pattern of afforestation has been influenced by differences in land tenure. In Scotland and parts of northern England, the sale of large estates has made big blocks of land available, whereas in parts of Wales the average size of farms tends to be smaller and new plantations form more of a patchwork. The spread of upland afforestation has been least in the Pennines, mainly owing to the prevalence of common grazing rights, and in some of the National Parks because of local opposition. Forestry has never been subject to planning control, but some local authorities have become increasingly concerned over planting proposals in sensitive areas. Under the existing consultation procedures new planting proposals are referred by FC to the Ministry of Agriculture, Fisheries and Food (MAFF), Welsh Office Agriculture Department (WOAD) and Department of Agriculture and Fisheries for Scotland (DAFS) for release from agriculture; to NCC (if the proposals are partly or wholly on an SSSI) for a nature conservation appraisal; and to National Parks Authorities (in National Parks) and local authorities (subject to agreement over consultation criteria between the relevant authority and FC) for an amenity appraisal. If there are objections from any of these authorities which cannot be resolved through discussion, the case is passed to the appropriate Regional Advisory Committee (RAC). There are seven RACs, appointed by and responsible to FC and consisting largely of members with forestry or land owning interests, but with representation of other country-side concerns. It is the RAC's role to act as an adviser to FC, but also to achieve a reconciliation between the parties in dispute. If disagreement still remains, the matter is referred to the Forestry Commissioners. If the Commissioners do not accept the objections to the planting proposals, FC always refers the matter to the appropriate Forestry Minister for his view. Technically the final decision rests with FC, but in practice a Minister's view prevails.

Forestry organisations include Timber Growers UK (TGUK), which is especially concerned with trade and

Afforestation policy and practice

marketing. The main professional body is the Institute of Chartered Foresters (ICF), but the Association of Professional Foresters (APF), Royal Forestry Society (RFS) and Royal Scottish Forestry Society (RSFS) are long-established groups with many practising foresters in their memberships. The main private companies are Economic Forestry Group, Fountain Forestry, Tilhill Forestry and the Scottish Woodland Owners' Association. The Forestry Commission, acting both as chief adviser to Government on national forestry policy and chief agent in the execution of this policy, has a close relationship with this non-state sector.

Forestry policy has been successful in meeting the objective of re-establishing forest cover in Britain, as 2·207 Mha (9·6%) of the country are now wooded (Forestry Commission 1985), compared with only 1·184 Mha of woodland or 5·3% of the country in 1924 (Commission survey). Allowing for an estimated 83,000 ha of existing forest lost to other land-use during the period (K. J. Kirby, unpublished), the increase in new forest since 1924 amounts to about 1·106 Mha. Despite short-term fluctuations, forestry policy has supported continuous expansion of new planting, which over the last 60-year period has averaged 18,133 ha per annum. The latest government forestry policy statement (Secretary of State for Scotland 1980) considered that new planting should continue at the rate of the last 25 years, that is 20-25,000 ha per annum.

Although new afforestation has mostly been restricted to soils low in fertility and agricultural value, it has had to cope with a wide range of environmental conditions, especially in climate and its inter-related effects on soil. Tree growth is favoured by sufficiency of rainfall combined with good soil drainage, high summer and winter temperatures, low exposure to wind, adequate soil depth and high soil nutrient status. Britain experiences marked gradients of climate which have a profound bearing on tree growth. Although the country as a whole has an oceanic climate, there is a great contrast between the relatively

continental conditions of low-lying East Anglia and the hyper-oceanic climate of the mountainous coastal districts of western Britain. This involves a tremendous increase in rainfall and atmospheric humidity (map 12) but also a decrease in temperature range. There is another important temperature gradient from the warm south of England to the cold north of Scotland (map 13). There is also a trend from prevalence of deep, well-drained soils in the eastern lowlands to shallow, waterlogged and peaty substrata overlying hard, acidic rocks in the mountainous west. The combined result is that conditions for tree growth (though good in the extreme south-west) become increasingly unfavourable in a north-westerly direction in Britain (see map in study by Centre for Agricultural Strategy 1980). A trend towards unfavourable conditions also occurs with increasing altitude on any mountain — increase in rainfall, cloud cover and wind exposure, and decrease in temperature, drainage, soil depth and fertility. These effects produce an altitudinal limit to the growth of trees, whether they are native or introduced species. The natural tree line has been almost totally lost through deforestation, but high-level remnants indicate that it would be at around 650m in southern and eastern mountains, descending to 300m or less in the far north-west of Scotland.

The original climax forest of Britain 3,000 years ago was mainly cool, temperate broadleaved woodland, variably composed of pedunculate and sessile oak, ash, wych elm, birch, alder, and small-leaved lime as the main trees. Only in parts of the Scottish Highlands was there an outlier of the Boreal coniferous forest zone of northern Europe, dominated by Scots pine as our major coniferous tree. In the far north of the Highlands there are also equivalents to the Subalpine birch woodland which lies mostly above the conifer zone in Norway. The native broadleaved trees are too slow growing and/or too inferior in timber qualities to be acceptable for economic afforestation. Conifers are regarded as the only commercially viable species, because

Afforestation policy and practice

they grow faster and several species perform well in cool, oceanic climates and on nutrient-poor, acidic soils, producing timber of adequate quality.

Some of the earliest FC plantations were on the deep podsolised sands and raw chalky soils of the East Anglian Breckland, where mostly Scots pine and Corsican pine were planted. Acidic heaths on the east Suffolk sandlings and in Dorset and some large coastal sand-dune systems have also been planted. However, the lower hill ground (i.e. the submontane zone) soon became the main source of plantable land. Norway spruce did well, but it was soon found that Sitka spruce from coastal British Columbia was ideally suited to the cool, oceanic conditions and the shallow podsols and wet gley soils so widespread in our uplands, and this rapidly became the favourite tree. Scots pine fell from favour, but Japanese or hybrid larch has been much planted as a reasonable performer on the drier sites in the hills. Grassy sheepwalks were the most favoured planting ground, but during recent years the uncertain financial fortunes of grouse management have made increasing areas of heather-covered moorland available for afforestation. Earlier planting was mainly on the drier or only moderately wet soils with shallow surface peat. Improved technology led to the introduction of ploughing as a standard ground preparation: not only did this improve drainage on wetter ground, but even on drier sites the upturned ridges gave the planted young trees a weed-free start. Eventually, it became possible to plough the deep peat of raised and blanket bogs to such a depth that water tables were lowered enough for trees to grow successfully even on these very wet sites, and during recent years these bogs have been increasingly used as a source of plantable land. Lodgepole pine in particular has been planted on these deep peats because of its tolerance of waterlogged soil.

The FC (1984) Census for 1979-82 shows that planting in the years 1971-80 was 60% Sitka spruce, 14% lodgepole pine, 5·5% Japanese or hybrid larch,

Breckland grass-heath on sand mixed with chalk and flint: habitat of the stone-curlew. Plantations of Scots pine and Corsican pine now cover 19,500 ha of the open brecks existing in 1920. Weeting Heath NNR, Norfolk, 1985.

Afforestation policy and practice

4·4% Scots pine, 10·7% other conifers and 5·3% broadleaves; this refers to all planting, including re-stocking, but most was new forest. The planting of intimate mixtures of lodgepole pine and Sitka spruce, as a self-thinning device in favour of the second species, will tend in time to increase the area of spruce-dominated forest. The ground to be planted is usually ploughed to a depth of 45-75cm depending on whether it has mineral soil or deep, wet peat, and the nursery-grown transplants or tubed tree seedlings are planted in the ridges upturned from the furrows, which are usually 2-4m apart. The use of double mouldboard ploughs on deep peat throws up twin ridges about 2m apart and trees are planted at 2·0-2·4m intervals along these. On less peaty sites, the latest technique is 'ripping' with a tine through the soil, which gives local drainage and a planting line. Fertiliser is applied on the poorer soils — usually rock phosphate (calcium phosphate), though on some soils potash is used as well. Urea or, occasionally, ammonium nitrate is also

used locally where soils are deficient in available nitrogen. Herbicides are used to suppress competing plants on the richer lowland sites and allow good establishment of the trees, but they are usually unnecessary on ploughed ground. Sheep are removed before planting begins, and boundary fences erected against their return from surrounding land. Attempts are made to reduce the numbers of large herbivores such as red deer, roe deer and feral goats by fencing and shooting. Networks of rides are left unplanted and dissect the forest into compartments, and systems of access roads are developed as the forest matures.

Research on tree nutrition and the selective breeding of improved varieties have enhanced tree performance, but afforestation in Britain often works against severe environmental limitations. During the last 15 years there has been an increased tendency to plant in unfavourable situations. Normal altitudinal limits to planting have generally been around 450m, rising to

A nationally important sand dune and slack system. The conifer forest in the background covers more than half the dune area existing in 1950. Newborough Warren, Anglesey, 1980.

Afforestation
policy and
practice

600m in Wales and falling to 300m in northern Scotland. These limits have been pushed ever higher in many districts, towards or even above the climatic tree line (e.g. 760m in Wales, 720m in Galloway). Planting of wet blanket bog 'flows' stops only where the ground becomes dissected into systems of pools or covered with *Sphagnum* lawn. Moreover, although the oceanic climate gives a favourable temperature regime for tree growth, an associated factor is wind, which not only reduces growth in exposed places but creates a high risk of the trees being blown down before maturity. Wind speed increases westwards and northwards, and with altitude, creating a hazard for forestry over much of western and northern Britain and in the hills generally (map 14). Conifer plantations on ploughed ground and on gley sand blanket peats seem particularly prone to windthrow. Various management techniques, including 'oceanic forestry', are being tried to counter the problem, but in general, once much of the crop is blown, the rest is felled. This is leading increasingly to stands in western and upland areas being felled at age 30-35 years, instead of at the optimum of 55-60 years in the normal rotation.

The outcome is that the yield of timber from the post-1919 forests is reduced and consists mainly of small sawlogs, small roundwood and pulpwood. This position is not likely to change unless low-lying and more fertile land becomes more extensively planted, which in addition to yielding more volume per hectare would also yield a higher proportion of sawlogs. A good market for the small-sized timber nevertheless exists through the development of wood-based panels and for pulp and paper products. Mills to cope with the increasing output of small roundwood have been established recently at strategic locations in Wales, northern England and Scotland.

The most recent government statement on forestry policy was in 1980. The salient points were: "there should be scope for new planting to continue in the immediate future at broadly the rate of the past 25 years while preserving an

The grassland sheep-walks of the southern uplands. These smooth, rounded and relatively fertile hills founded the Tweed wool industry. There are numerous planted shelterbelts of conifers, but fragments of native broadleaved forest survive in rocky glens. Meggat Water, Tweeddale, Selkirkshire, 1972.

Afforestation policy and practice

acceptable balance with agriculture, the environment and other interests. We see a greater place for participation by the private sector in new planting, but the Forestry Commission will also continue to have a programme of new planting." (Secretary of State for Scotland 1980). This has to be set against the option recommended by the Forestry Commission, in *The wood production outlook in Britain* (1977), for planting up to another 1·8 million hectares of new forest, and against the case made by the Centre for Agricultural Strategy (CAS) in *Strategy for the UK forest industry* (1980) for planting up to another 2·0 Mha. The CAS study contained a map and tables (in Appendix IX), based on FC data, showing the gross areas of rough grazing in Britain which could be considered for forest. Within a total area of 6·145 Mha (which excluded the Outer Hebrides and large areas of the north-west Highlands, presumably on grounds of climatic severity), 1·025 Mha were already in forest and 2·140 Mha were unsuitable (too high, too steep,

rocky, undrainable, or occupied by roads and buildings) or inappropriate (the better agricultural land). This left 2·980 Mha as technically suitable for afforestation, but, when constraints of water supply and scenic amenity (though not nature conservation *per se*) were taken into account, only 1·1 Mha were free of constraint. Table 1 has revised these figures.

The total extent of land which the forestry interest regards as plantable is, in practice, somewhat flexible. Constraints are by no means regarded as constituting a ban on afforestation, and there has been a reluctance to give undertakings in respect of any constraints. The CAS amenity constraint included all the National Scenic Areas in Scotland (1·0 Mha, but not all plantable), but these are regarded as areas where "forests can increase, provide more diversity while still retaining the overall character" (Balfour 1981). Proposals for formal controls on afforestation within National Parks have also been successfully resisted (see p.68). Changes which will make

The dry heather-clad grouse moors of the eastern Highlands, with coniferous shelterbelt. Bearberry (dark green) grows in patches amongst the heather. Classic habitat of the hen harrier and merlin, as well as red grouse, these moorlands have been increasingly afforested since 1970. Tornahaish, Strathdon, Aberdeenshire, 1981.

3

**Nature
conservation
and afforestation**

**Afforestation
policy and
practice**

upland commons available for afforestation have been advocated (CAS 1980). And greater freedom has recently been achieved in Scotland from the major constraint of veto by the Agriculture Department on release of hill and marginal land from agricultural use. The afforestation of more fertile areas in the agricultural lowlands is also being widely urged by various interests as a means of alleviating the problem of agricultural over-production. The definition of plantable land is also periodically revised, in the light of new establishment techniques and socio-economic factors. Afforestation has since occurred or been proposed within areas not shown as suitable on the CAS map (e.g. Islay, Wester Ross and West Sutherland, and at higher levels on Skye). While no targets for an eventual planting area have been set within national forestry policy, private forestry interests have strongly supported the top planting options of the FC and CAS studies (e.g. Crawford 1979), and some foresters are confidently asserting that a doubling of

the present forest area will be achieved by 2025 AD (e.g. Ogilvie & Lamb 1986).

Within its forests the Forestry Commission has shown concern for wildlife management and conservation, and in the Countryside Acts of 1967 and 1968 it was required to "have regard to the desirability of conserving the natural beauty and amenity of the countryside," a formulation which included nature conservation. There was also a specific duty to protect against polluting any water used by statutory water undertakers. During the debates in 1985 on the Wildlife and Countryside (Amendment) Bill, an attempt to add a responsibility to further conservation to FC's terms of reference was resisted by the Ministers concerned and replaced by a requirement to endeavour to achieve a reasonable balance between
a) the development of afforestation, the management of forests and the production and supply of timber, and
b) the conservation and enhancement of natural beauty and the conservation of flora, fauna and geological or physio-graphical features of special interest.

Ploughing of blanket bog and wet grassland over gley soils, in preparation for planting. Water tables are lowered and run-off and erosion increased. Spadeadam Forest, Cheviots, Cumbria, 1983.

Environmental changes caused by afforestation

Afforestation causes profound changes to both the living and the non-living components of an ecosystem. It is a fundamental change in land-use, involving an accelerated succession from open ground to forest, with conspicuous alterations to wild plant and animal communities and less visible modifications to the physical and chemical conditions of soil and water. The complexity and variability of these changes make it difficult to give a short yet adequate description of them. Afforestation has affected a wide range of environments and ecosystems spread over a large part of Britain. Moreover, on the assumption that forestry will continue indefinitely on the afforested areas, the forests will go through repeated cycles of harvesting, re-stocking and re-development which will give opportunity for planned adjustments to their character. The present account is limited to a brief summary, and readers interested in further detail should consult the referenced sources.

Changes vary according to differences in geographical location and climate, habitat type and management/silvicultural regime. The last category includes not only the techniques and tree species used, but also the extent and type of open ground which is left within the forest and adjoining it. All new forests have a minimum extent of residual open ground in the form of access roads and rides, amounting to about 5%. In areas where virtually all the land is plantable (e.g. lowland heaths) further open areas will tend to be small, consisting of deer-lawns, recreational sites, power line corridors, some archaeological sites, and ground unplanted for landscaping purposes. In low hill and moorland country the same may be true, though areas of more fertile land, such as enclosed pastures around occupied habitations and the banks of larger streams, are often left. In many forests, patches of lower-lying spongy bog and more extensive higher plateau blanket bogs have been left unplanted. Cliffs and screes are mostly unplantable, and rocky hills with outcrops, pavement and thin soil are usually left unplanted. The higher mountains, with a large extent of montane ground well above the tree line, can carry only a lower zone of forest, as in many parts of the Highlands. Sometimes trees remain stunted ('checked') or die within certain parts of a forest; these 'checked' and failed areas may be fertilised or replanted, but in some cases they are left and form open enclaves. In general, most of the land acquired by forestry concerns below the altitudinal planting limits is planted with trees, and residual open ground is usually a quite minor element of the new forest. The most important changes are thus the successional developments imposed by the trees themselves. Remaining open ground habitats within the forest often show change, too, and this is dealt with in the next chapter (see p.50). Ratcliffe (1986) has also made a more detailed assessment for Scottish forests.

Effects on land plants and animals

During the first few years after planting, the young trees form no more than rows of small bushes and the pre-existing vegetation remains dominant. Under relaxation from grazing and burning, the grassland, heath, bog and sand-dune species all become taller and more luxuriant. Grasses and their relatives (e.g. purple moor-grass, mat-grass, fescues, bents, cottongrasses, heath rush, deergrass and various sedges) grow dense and tussocky, and a litter of their dead leaves accumulates. There is a still more dramatic effect on dwarf shrubs such as heather and bilberry, especially where these had become poor remnants under heavy grazing: they grow tall and begin to spread at the expense of the grasses. Where dwarf shrubs had previously been eliminated, they do not return, and a dense grassland prevails. The effect on herbs other than grasses and sedges varies according to their size: small species may be overwhelmed by competitors, but larger ones tend to grow taller and flower more freely than before. Mosses and lichens often form deeper and more luxuriant carpets. Plants needing really wet conditions tend to dwindle or disappear rapidly because of the drying effect of ploughing; examples

Environmental
changes caused by
afforestation

are sundews, butterwort, bog bean, bladderworts, sedges and bog-mosses.

The effects on animals are equally dramatic. Some birds of open ground disappear at once, but others may linger for a few years before being displaced by the growing trees. Species discouraged by the young trees are stone-curlew, ringed plover, stock dove, lapwing, wheatear and skylark in the Breckland (Clarke 1937; Lack 1939), and the last three species and also golden plover, dunlin, snipe, greenshank, redshank, curlew, red grouse and meadow pipit on the northern moorlands and sheepwalks (Reed 1982). Some open ground songbirds are favoured and may increase until the trees close, for example whinchat, stonechat, grasshopper warbler and woodlark (in Breckland), and species of open woodland, scrub or woodland edge typically colonise the pre-thicket stages, for example tree pipit, willow warbler, whitethroat and black grouse (Lack 1933; Reed 1982; Harris 1983). Dartford warblers increased

temporarily on afforested heaths in Dorset and Hampshire but disappeared at the thicket stage (C. R. Tubbs, unpublished).

Increased amount of vegetation and litter provides a boost in food and cover for some animals, notably the short-tailed field vole, which often shows enormous temporary increases in abundance. Voles are the prey of various bird and mammal predators, and their abundance attracts open country species such as short-eared owl (Goddard 1935), kestrel (Village 1982) and, more locally, hen harrier (Watson 1977). However, large bird predators sooner or later find their food supply depleted and leave. Ravens are so dependent on sheep carrion that they usually cease nesting at once and then gradually move away as the trees close in and further restrict their search for food (Marquiss, Newton & Ratcliffe 1978). The golden eagle may benefit for a few years, but as the forest closes and the supply of grouse, hares and rabbits become unavailable, breeding success declines and territory desertion

Recently established Sitka spruce forest. Increased luxuriance of purple moor-grass and heather allows field voles to increase and attract predators such as short-eared owls. Black game increase at this stage. Big Corlae, Water of Ken, Galloway, 1978.

Environmental changes caused by afforestation

eventually occurs (Marquiss, Ratcliffe & Roxburgh 1985). The same is true of the common buzzard in some areas, for example in the Galloway hills (D. A. Ratcliffe & R. Roxburgh, unpublished). Merlins may linger for some time, but eventually disappear from extensive areas of closed forest (A. D. Watson unpublished; Newton, Meek & Little 1978). Peregrines are unaffected in areas where their principal prey is domestic pigeons passing over the forests, but they must be vulnerable in parts of the Highlands where they depend largely on wild prey of open ground (Ratcliffe 1980).

The abundance of voles, other rodents and small birds encourages stoats and weasels, whilst polecats do well in young forest in Wales, and pine marten and wildcat in parts of Scotland (Staines 1983). Foxes feed on these small prey species as a substitute for sheep carrion and larger birds or mammals. Both red deer and roe deer usually manage to enter young forests, despite fencing, and benefit from a copious growth of vegetation. Rabbits, brown

hares and mountain hares sooner or later decline. Other animals, including insects, are necessarily affected. Decline of the large heath butterfly and moths such as the emperor, fox and northern eggar are examples within the Lepidoptera, but there are few studies of effects on the open ground invertebrate communities.

During the first ten years, the young forest is usually good wildlife habitat, with features otherwise unusual in Britain. Although some of the open ground features are rapidly lost, there is a type of ecosystem which contains elements from both treeless ground and open scrub. Then, as the young trees expand and coalesce, their dense growth increasingly suppresses all other vegetation; and by the time the young plantation closes to form thicket, at 10-15 years, other plant life has almost totally disappeared within the planted areas, and much of the previous animal community with it. Subsequent developments depend on location and silvicultural practice. Tree density in the maturing forest is all-important,

Well established Sitka spruce forest at the thicket stage. Much upland forest is grown in dense stands until felling and contains little vegetation other than in rides and along edges. It is attractive to songbirds and the cones attract crossbills and siskins. Polmaddy Glen, Carsphairn, Galloway, 1985.

4

**Nature
conservation
and afforestation**

**Environmental
changes caused by
afforestation**

because it determines light penetration, which in turn affects the development of vegetation under the trees, and hence the associated animal communities. Failure to brash and thin the young trees results in a dense, almost impenetrable and intensely dark thicket woodland in which plant life, apart from soil organisms, is almost totally suppressed. Not even shade-tolerant ferns, mosses and liverworts can survive. Woodland birds colonise, but are largely confined to the canopy and edges, and consist of species able to subsist under these limited woodland conditions, for example goldcrest, coal tit, chaffinch, robin, wren, song thrush, blackbird and woodpigeon. A community dominated by songbirds replaces one in which waders are especially well represented (Moss 1978; Reed 1982; Harris 1983). Rarer songbirds which appear locally or periodically, according to variations in seed production, are crossbill and siskin (Newton 1972). There may be shelter for some of the mammals, but herbivores are especially limited by the poor food

supply (Hibberd 1985). Light penetration varies according to the tree species, but the thinning regime is crucial: spruce gives especially deep shade, but even larch and Scots pine grown in dense canopy can virtually eliminate the ground flora.

As brashing and thinning increase, and as the forest matures, conditions for wildlife within the tree compartments improve steadily. Ground vegetation reappears in close relation to the degree of increase in light penetration. In well-thinned but dark sprucewood there is often a moderately dense growth of mosses but only a patchy cover of grasses, bilberry and ferns. In well-thinned pine and larch woods, dense growths of grass and ferns (especially bracken and buckler-fern) often dominate, with much bramble in lowland situations (Hill 1979, 1983). The flora of conifer plantations is usually very limited, but rare species such as creeping lady's-tresses and one-flowered wintergreen have locally colonised plantations of Scots pine. There may be local luxuriance of

Well thinned plantations of Scots pine, allowing good light penetration and development of a continuous field layer of bracken and grasses. Thetford Chase, Breckland, Norfolk, 1981.

Environmental changes caused by afforestation

ground mosses and lichens, but species diversity is limited and conifers usually have few of the bark-growing lichens and mosses which are so characteristic and important on native broadleaved trees (P. W. James, unpublished). Thinning also enhances opportunities for birds, especially larger species such as woodcock, carrion crow, tawny owl, long-eared owl and sparrowhawk. The goshawk is, very locally, one of the most important additions to the new forests, and, though its presence is attributable to deliberate introductions (Marquiss & Newton 1982), it is now spreading spontaneously. Variety in birds is almost invariably limited by the general absence of a woody shrub layer and by the scarcity of dead timber which can be used by cavity-nesting species. Provision of nest boxes may alleviate the latter defect for songbirds (Currie & Bamford 1982), but dead and dying wood is also necessary as feeding habitat for species such as woodpeckers. Abundance of ground vegetation greatly increases carrying capacity for herbivorous mammals,

from small rodents to deer, and thus enhances the habitat value for carnivores. Conifer woods, especially of pine, have become important habitats for the red squirrel in certain districts, such as Breckland.

The extensive coniferous plantations have encouraged a substantial arboreal insect fauna, composed both of species originally widespread on native conifers and also of others which have greatly increased and spread because of the abundance of new habitat. One of the most notable in the second category is the pine hawkmoth, now well established in the pine forests of Breckland and elsewhere in the south and east of England. Some of the native insects have, however, become significant pests in certain areas (see p. 42). Conifer wood interiors — as distinct from rides — are seldom butterfly habitats, but the speckled wood butterfly has become common through most of the Breckland forests.

The next important factor is the age at which the forest is felled. This depends on soil quality, tree species,

Windblown young pole stand of Sitka spruce. The blown area extends to the standing forest edge. Windthrow problems lead to a lack of thinning and early felling, producing thicket plantations and short rotations.
Kielder Forest, Northumberland, 1986.

Environmental changes caused by afforestation

regional climate and altitude. Relatively little conifer forest is allowed to grow to full maturity and the normal felling rotation in state forests has been 55 years (Forestry Commission 1977). Clear-felling is the norm, and in western and upland districts it has been taking place at an earlier age to circumvent windthrow. While the risk of windthrow increases with age and size of tree, it typically occurs after the first thinning, leaving holes in the forest varying from a few square metres to several hectares. The larger windthrown areas are cleared and replanted to give an unplanned extra element of age-class diversity. After normal felling and timber extraction, the cut branches and foliage ('slash') from the trees are usually left and the area is replanted, typically with the same tree species.

After restocking, ground vegetation usually increases spontaneously again, especially where it has been sparse or absent, but the original pre-afforestation communities are not restored. Some of the species previously present reappear, springing up from buried seed, from vegetative growth persisting within the forest phase, or by colonising from sources outside (Hill 1979, 1983). But there is also a group of new colonists, mainly widespread and freely spreading species of open disturbed ground associated with human activity. The vegetation which develops is thus a mixed type, but varying according to geographical and ecological situation. Previous open ground species include gorse, bracken, heather, bilberry, cottongrass, heath rush, tormentil, heath bedstraw, purple moor-grass, bents, wavy hair-grass, tufted hair-grass and common pleurocarpous mosses. Species typical of woodland clearings or disturbed ground include rosebay willowherb, foxglove, bramble, buckler-fern, male-fern, hard fern, wood-rushes and fog grasses. The community often becomes luxuriant and dense, but where large amounts of tree litter remain they reduce the cover of the redeveloping vegetation. When such debris is burned or the soil disturbed again in preparation for planting,

Clear-felled and restocked area of forest. Grassy vegetation is already redeveloping and will form a dense, tussocky cover, mixed with the lop and top, providing good habitat for woodland edge birds. Kielder Forest, Northumberland, 1986.

Environmental changes caused by afforestation

acrocarpous mosses often become abundant. Because the afforestation process tends to dry out previously wet ground, a noticeable feature is that the resulting second rotation ground vegetation is usually poor in plants of moist and boggy habitats, apart from certain *Sphagnum* mosses typical of · damp woodland. Except where old ditches form waterlogged channels, peatland species do not reappear and the redevelopment of whole mire communities is not possible. To the extent that many pre-afforestation upland sites had a range of conditions from dry mineral soils to deep, wet peats, there is inevitably a loss of ecological diversity in the open ground communities of the second and subsequent rotations. Hill (1979) has also suggested that there may be a progressive loss of open ground species through succeeding rotations.

This renewed open ground vegetation which develops within a few years, including the young trees, is colonised by birds typical of such habitat. A good variety of breeding species often becomes established, but tends to decrease with altitude, lowering of soil fertility and distance northwards. Bibby, Phillips & Seddon (1985) reported up to 31 nesting species in restocked conifer forest on some more fertile, lower-level sites in Wales, whereas Leslie (1981) found 12 species on poorer, higher-level sites in northern England. This bird community is typical of open woodland and scrub or dense grassland and usually has poor representation of species from the previous open heath or moorland. Songbirds predominate and waders are few, for the often dense regrowth of vegetation and the persistence of slash and of furrows from the original ploughing make this an inhospitable habitat for birds whose chicks leave the nest soon after hatching. In North Wales, Currie & Bamford (1981) found that restocked forest plots had a much greater diversity and density of breeding songbirds than newly afforested plots on similar ground, mainly because of the amount of residual woody growth and forest edge on the restocked sites. Restocked forest has become important habitat for the

nightjar in England and Wales, and Gribble (1983) found that almost half the birds recorded in the 1981 national survey were associated with conifer woodland, though he noted that commercial forestry is a mixed blessing for nightjars, because of its overall effect on open habitat loss. At least a fifth of the British population of the woodlark, another mainly heathland species, is now estimated to occur in Breckland, all nesting here on restocked forest sites in 1984 (Leslie & Hoblyn 1984). Herbivores benefit from the renewal of a herbaceous or shrubby vegetation on these open areas; as they become more extensive, the feeding value of the forest increases for deer in particular, though this can give problems for tree re-establishment (Hibberd 1985). Bird and mammal predators within the established forest will also gain appreciably from the enhanced food supply and could conceivably increase in density, compared with the first generation forest. Open ground invertebrates, including butterflies, have a boost again but have not been studied in detail. The 0-10 year restocked phase is a distinctive habitat in its own right, combining elements of woodland edge, glade, scrub and open ground, but it is rapidly changing and temporary.

Once the growth of young trees closes again, after ten years or so, the process of the first rotation is repeated, in that a wholly woodland community of plants and animals becomes re-established within the replanted areas. Whatever the prospects for silvicultural modification of the second and subsequent generations (see pp.46-49), they will remain essentially forest ecosystems during the maturation phases. The long-term possibilities for retention of open ground wildlife in the new forests will thus depend on the extent, configuration and ecological character of unplanted habitats and restocked ground in the 0-10 year age class. The proximity and extent of unplanted open ground outside the forest will also remain important to some of the larger herbivores and predators which are limited by the feeding value of the forest area itself (e.g. deer, fox, carrion crow, buzzard, sparrowhawk and goshawk). The open ground phase represented by felling and restocking will necessarily have a dynamic character over time, continually shifting around the whole forest area and changing in spatial pattern. This shifting pattern should give interesting opportunities for ecological study, including comparisons with more static open ground habitats both within and outside the forest. It remains to be seen whether some kind of steady state will eventually develop in this dynamic pattern, but it is obvious that, as forest rotation period decreases, the relative extent of open 0-10 year phase at any one time must increase while opportunities for diversification within the 10 year to felling phase become less, and *vice versa*.

An important but somewhat neglected aspect of afforestation is the ecological significance of the relationship between the forest area as a whole and unafforested ground outside it. This is especially important for the more mobile animals which can range between the two habitats, but it applies also to the more readily dispersing plants, including trees. A key group of species here is the group of larger predators which find shelter and breeding sites amongst the trees, but either need substantial areas of open ground for feeding or find thicket plantation unsuitable because of its low food value or the difficulty of hunting within it. Where there is a large interface between open ground and forest, some such species may flourish, reaching high breeding densities within the trees, but this situation is likely to depend on the continued existence of large areas of open ground beyond. The prospects for species such as common buzzard, red kite, goshawk and even sparrowhawk in new conifer forests would seem sensitive to this factor, but more needs to be learned about food requirements and carrying capacities of different habitats.

Another important issue is the effect of the forest-dwelling predators on prey species outside. A benign effect of afforestation is the general

protection afforded to most predators, some of which are still illegally persecuted on certain open areas beyond. FC, in particular, has made considerable efforts, notably through its ranger service, to look after legally protected species. Others regarded as pests, especially crows and foxes, are controlled, but in large areas of thicket forest this can be physically very difficult. Overall, afforestation generally results in a relaxation or even an absence of traditional keepering activity. There is growing evidence that this predator force is depressing both the breeding success and the numbers of at least some bird species on the adjoining moorlands. Such effects are reported on both red grouse (R. van Oss, unpublished) and golden plover (R. Parr, unpublished). Because of the ranging capacity of foxes and crows, this edge effect could extend over a zone up to several kilometres wide. Some moorland nesting waders appear to avoid the immediate vicinity of large conifer blocks with wall-like edges, though they will nest closer to more open areas of native pinewood. Stroud & Reed (1986) found that curlew, golden plover, dunlin, lapwing and redshank showed avoidance within 400m of forest/moorland boundaries: the pattern persisted up to at least 800m, but less markedly for curlew. This avoidance effect is likely to have a marked influence on the minimal area of open moorland needed by these species for nesting. The competitive effect of sparrowhawks has also been suggested as an additional reason for the decline of the merlin in some afforested areas (D. N. Weir, R. Dennis, unpublished).

The significance of the relationship between the two habitats for deer is evidently complex and variable according to species. Management of deer is important in the new forests, because of their capacity for causing damage both to the trees (by browsing leading shoots of young saplings and bark-stripping at various ages) and to land and crops outside. They also provide a crop and source of sporting revenue in themselves. Red deer which can move freely between extensive

Environmental changes caused by afforestation

adjoining open ground and the forest are not limited by food supply and can build up to large numbers, with considerable potential to become pests in both directions if they are not controlled. Roe deer tend to be more confined to forest areas, but in places they are increasingly adapting to open moorland. Feral goats can do considerable damage within new forests, and foresters often need to control them.

In the interchange of plant species, the capacity of conifers for natural regeneration on habitats outside the forest is only very locally a problem so far. However, on Lakenheath Warren in the Breckland, Scots pine from adjoining plantations has regenerated so freely on remaining areas of short calcareous grassland (a type with high nature conservation value) that the resulting young woodland has been dedicated by the estate. On some sand-dune systems, regeneration of conifers from adjoining plantations onto the open dunes is also causing conservation problems. Habitats outside the forest serve to a limited extent as parent

sources of propagules for the spontaneous regrowth of ground vegetation on restocked areas.

Effects on hydrology

The extensive draining operations integral to afforestation of moorland and bog have immediate effects on hydrology, causing increased rate of run-off from the ground and flow in the receiving streams. The effect is exacerbated by the general practice of ploughing at right angles to the contours. This is entirely to be expected on the wetter types of ground, since a main purpose of ploughing here is to remove standing water and lower water tables. There follows a marked tendency to wet weather river spates followed by rapid subsidence to low water flows since the timing and amplitude of run-off from the catchment are disturbed. Many small streams, springs, rills and flushes are completely destroyed and converted into drainage runnels by the preparatory ploughing. In extensively afforested catchments, such as those of the Wye and Severn in Wales and the Fleet and Cree in

The pattern of ploughing for afforestation on hill ground. Such drastic treatment inevitably increases water run-off and soil erosion, and reduces the value of the habitat for some moorland birds. A quarry has been excavated to supply road-metal. Polmaddy Glen, Carsphairn, Galloway, 1985.

Environmental changes caused by afforestation

Galloway, the tendency to 'flash' spates has become well known. Robinson (1980) showed that in the first seven years after ploughing a Cheviot catchment for afforestation, there were higher flood flows and the time to stream peaks was halved, while annual water yield was increased by 5%. The effects of increase in rate and amplitude of stream flow on cave and karst sites of geological value need to be examined. Blanket peat dries irreversibly after afforestation, and the development of large shrinkage cracks under lodgepole pine in northern Scotland is reported (Pyatt & Craven 1979).

In theory, as the forests mature and create a greater water uptake demand, the system should stabilise, so that river flows become evened out again and flash spates reduced. Relevant information to establish a general picture has proved difficult to obtain, but Robinson & Newson (1986) have shown that faster flood responses still occur from the mature forested catchment at Plynlimon than from open moorland. The issue has also become complicated by the indisputable evidence that large blocks of conifer forest actually reduce the amount of water running off the area, mainly through interception losses (Law 1956; Calder & Newson 1979). Continuing work by the Institute of Hydrology has shown that these interception losses increase with rainfall. On Plynlimon (mean annual rainfall 2,350mm) the proportion of rainfall reaching the rivers was 85% on the unafforested grassland sheepwalks draining to the Wye, compared with 65% on the largely afforested catchment of the Severn (Newson 1985). Typically, a decrease of run-off of about 20% is also predicted as a result of afforestation in Scotland (Calder & Newson 1980). Reduction in water yield through afforestation is believed to be less when heather- or bracken-covered ground is planted, because these types of vegetation themselves cause greater interception losses than grassland, and this is currently under investigation. The interception effect is quite small in the Breckland (Gash & Stewart 1977), but

most of the recent planting has been in high rainfall areas of the north and west. Other types of woodland might also be presumed to have a comparable effect, but in the United States Swank & Douglass (1974) reported a 20% reduction in water yield 15 years after two catchments had been converted from mature deciduous hardwood to white pine. This reduction was considered to be caused by the greater leaf area of pines, which intercepted and evaporated more rainfall than did deciduous hardwoods. On geological formations lacking rock aquifers this effect could increase the propensity for streams to dry up during critical drought periods, and may also affect the rate of recharging of depleted aquifers. When the trees are clear-felled, there is another increase in run-off, but if the forest can be managed by the cutting of a similar area each year, flows from the whole area should be evened out (Binns 1979).

Effects on sedimentation and erosion

Another increasingly recognised effect is on sediment transport in run-off water. The extensive draining of catchments, and especially the vertical ploughing of slopes, produces an increased downwash of mineral and organic sediment, mainly from the erosion of the furrows, which can cover up to 20% of the ground surface. Moreover, the increased spate of streams creates greater erosive power along their courses, causing both an increment in transported material and a redistribution of existing sediment deposits. Robinson & Blyth (1982) found that the average sediment loads increased at least 50-fold in an affected stream during draining operations and took several years to decline to a new equilibrium, which was four times higher than before drains were cut. An estimated 120 tonnes/km of sediment were lost from the catchment during the five-year observation period as a result of draining, compared with an estimated 15 tonnes/km for the five years before draining. On Plynlimon an afforested catchment continued to produce four times more stream sediment bedload than a pasture catchment, and ditches on steep slopes

Nature conservation and afforestation

4

Environmental changes caused by afforestation

Species displaced by new forests
1 Stone-curlew
2 Golden plover
3 Golden eagle
4 Raven
5 Red grouse
6 Greenshank

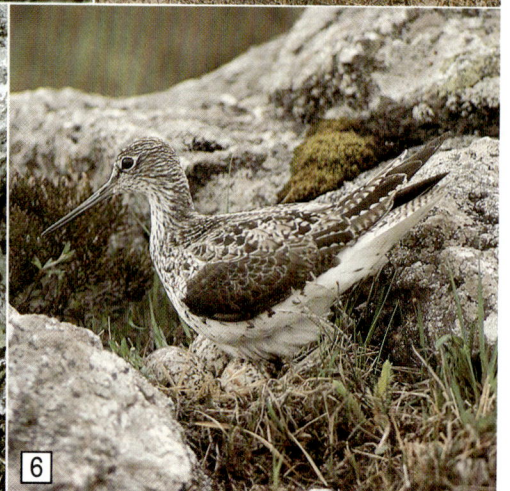

36

Environmental changes caused by afforestation

37

Environmental changes caused by afforestation

were vulnerable to erosion to the bedrock (Clarke & McCulloch 1979).

Battarbee *and others* (1985) compared lakes in unafforested and recently afforested catchments in Galloway. Three unafforested lakes showed little change in sedimentation, but in four lakes with 15-70% afforestation of their catchments there was a marked increase in sediment accumulation coinciding with afforestation ploughing and resulting from increased rate of soil erosion in the catchments. In Loch Grannoch (70% afforestation), accumulation increased from 0·1 cm/yr to over 2·0 cm/yr during the disturbance period, but at Loch Skerrow (15% afforestation — see cover photograph) the rate declined to pre-disturbance level after about ten years. After partial afforestation of the Cray Reservoir catchment in South Wales, heavy rainfall caused an enormous increase of sediment loads in inflowing stream water: the peaks were during heavy rain, but shortly after this ceased one stream showed turbidity 44 times higher and suspended solids 93 times higher at a point below the forest compared with one above it (Stretton 1984). Binns (1979) notes that the additional disturbances of road-making and clear-felling also produce an increase in suspended solids in run-off water. Windthrow often produces great soil disturbance, throwing up banks amongst the roots of fallen trees. Cave sediments could also be subject, in different places, to increased erosion and to changed deposition.

Effects on soil and water chemistry

There has been much debate over whether or not conifers have an adverse effect on their soils. There is evidence that the same tree genus, including broadleaves such as beech, can form mull or mor humus depending on the soil type. Growth on poor soils leads to higher polyphenol composition, reduced litter decay, mor humus accumulation and further soil deterioration (Coulson, Davies & Lewis 1960; Davies, Coulson & Lewis 1964). The balance of evidence from numerous studies is that conifers growing on poor soils promote greater surface organic matter accumulation, greater acidity

and a higher degree of podsolisation (Miles 1978, 1986); so this is likely to be the prevailing situation in most British conifer plantations. Andersen (1969) showed that, in the pollen and chemical record of the previous interglacials in Denmark, the rise to dominance of spruce was consistently associated with a change from prevalence of mull to mor humus and associated plant communities. Miles (1986) suggests that the acidifying effect of increased organic layer accumulation under conifers results from increased base exchange capacity, which on nutrient-poor soils leads to relatively greater addition of hydrogen ions. He found that organic content and acidity reached a peak and then declined again, as the forest matured, but did not revert to the former level before planting. Many northern British forests are being felled well short of maturity, when organic content and acidity are high. Miles also found that the depth of the eluviated layer of podsols increased under Sitka spruce. Pearsall (1938) found that the oxidation processes accompanying drying of acidic peats caused a further marked rise in acidity, a combined effect to be expected from afforestation.

By contrast, certain broadleaved trees, notably birch, aspen and holly, can reverse the process found under conifers, leading in the direction of mull humus formation and increased diversity of plant species in the field layer (various authors; reviewed by Miles 1986). Ericaceous shrubs such as heather and bilberry have similar acidifying effects to conifers, whereas bracken and bent-fescue grassland are improvers (Miles 1986).

The situation has become complicated by the effects of acid deposition arising from urban-industrial air pollution, which began with the Industrial Revolution (Flower & Battarbee 1983). Increased levels of water acidity in lakes and rivers follow a geographical-geological pattern, being especially high in districts where high atmospheric sulphur dioxide levels coincide with prevalence of hard and acidic unbuffered rocks and soils (Fry & Cooke 1984). There is now evidence that extensive conifer forests further

4

**Nature
conservation
and afforestation**

**Environmental
changes caused by
afforestation**

elevate the levels of water acidity within
their catchments, mainly through the
efficiency of their mass of evergreen
foliage in trapping acidic aerosol
particles during lateral air movement.
In the Loch Ard area, 30km north of
Glasgow, Harriman & Morrison (1982)
found that water in streams from
afforested catchments was consistently
more acid than in streams from
unafforested catchments (mean pH
4·53 against 5·32) and had higher
concentrations of aluminium and
manganese (under low pH there is
increased mobilisation of metallic ions
held in the soil exchange complex). In
Galloway, Harriman & Wells (1985)
reported that stream water in a 60%
afforested catchment had mean pH
4·32 compared with 5·20 for a stream in
an adjacent, unafforested catchment —
an eight-fold increase in acidity. In
Wales, streams draining afforested
upland have also been found to have
higher acidity than those draining
comparable unafforested areas both in
the upper River Tywi catchment (pH
range 4·7-5·4 against 5·0-5·6: Stoner,
Gee & Wade 1984) and in the
headwaters of the Severn and Wye on
Plynlimon (pH mean 5·3 against 6·7:
Newson 1985). Rainfall acidity in the
Tywi catchment was relatively low, and
pH differences here were smaller than
in south-west Scotland, though still
appreciable.

This 'scavenging' effect of conifer
forest on water acidity should thus
decline if atmospheric pollution is
substantially reduced. Stoner, Gee &
Wade (1984) suggest that, on theoretical
grounds, forest acidification of soils is
unlikely to have a further effect on
waters and that precipitation acidity is
unlikely to acidify soils, but empirical
evidence on these aspects is needed.
Acidification effects of various kinds are
reduced on base-rich soils. In North
America, the classical Hubbard Brook
ecosystem experimental studies
showed that cutting of a forest (without
removal of timber) caused massive loss
of nutrients into run-off water (e.g.
Bormann *and others* 1974). Roberts
(1985), reviewing nutrient losses from
upland catchments, regards this as an
unusual situation and suggests that

losses through normal forestry practice
are usually less, though variable, and
can be reduced through appropriate
practices. The chemistry of the forest
ecosystem in Britain is further
complicated by the general practice of
adding fertiliser to boost tree growth,
thereby modifying nutrient budgets
and cycling, obscuring the previous
chemical picture and overriding some
of the natural limitations of soil. The
effects on soil are, however, apparently
less known than those on waters.

The application of ground rock
phosphate and other fertilisers to
improve the nutrient status of afforested
ground and promote tree growth has
the immediate effect of enriching
drainage water (Binns 1979). There is a
rapid initial loss of phosphorus from
fertilised catchments in stream water,
with winter highs thereafter (Harriman
1978). Loss in Loch Ard Forest continued for at least 3 years after application,
by which time it had reached 15% of the
total phosphorus applied. Loss of
potassium over the same period was
20% of the amount applied. Total
nitrogen loss was less but varied from
4% at Loch Ard to 9% in Braes of Angus
Forest. Under certain conditions,
fertiliser loss can be enormous. When
rock phosphate was applied to forests
in the Cree catchment in Galloway
during frosty weather, ensuing thaw
and rain produced a river load of 321kg
of elemental phosphorus in one day,
25 times the normal level (Coy 1979).
Eutrophication effects, linked
especially to phosphorus accumulation,
have been reported in some streams,
but even more for lakes and reservoirs
(see p.41). The calcium oxide
component of rock phosphate should,
in theory, resist the tendency to
increased acidification under conifers
and should reduce water softness, but
this aspect seems not to have been
studied. Flower & Battarbee (1983)
showed, however, that the trend of
increasing acidity in Loch Grannoch,
Galloway, was only temporarily
arrested by the physical and chemical
disturbance of forest establishment,
and that decline to lower pH levels
continued again subsequently.
Enhanced release of nutrients resulting

Environmental changes caused by afforestation

from disturbance to the ecosystem from clear-felling is likely also to cause some temporary enrichment of stream and lake waters.

Effects on aquatic plants and animals

High water acidity is harmful to fish, but the relevant issue here is whether the enhancement of acid deposition by conifer forests has a further effect. Drakeford (1979, 1982) has correlated substantial reduction in catches of salmonid fish in the River Fleet, Galloway, with expansion of afforestation in the catchment. Rod-caught salmon in 1971-78 averaged only one third of numbers taken during 1960-70, while average sea trout catches during 1972-78 declined to less than one tenth of those during 1960-71. In this district every major river catchment has been subject to substantial afforesta-tion, so that there are no 'control' systems in which to study the effects of acid deposition uncomplicated by this additional factor. Such comparisons have, however, been made elsewhere. Harriman & Morrison (1982) found that mayfly larvae were sparse or absent in forest streams in the Loch Ard area but plentiful in adjoining open moorland streams. Brown trout were present in only one stream draining a forest catch-ment, but occurred in all the open moorland streams. In the upper Tywi, Stoner, Gee & Wade (1984) found that afforested streams had no trout and an invertebrate fauna of only 23-27 taxa, whereas trout were present in all but one of the comparable unafforested streams, which had 46-63 invertebrate taxa. On Plynlimon, Newson (1985) found that trout had an average occurrence of 0·9 per 100m of afforested stream, compared with 11·1 per 100m of unafforested stream. Declines of salmonid fish in afforested catchments are reported in other studies (e.g. Graesser 1979).

Harriman & Morrison (1982) planted fertilised salmon eggs in Loch Ard streams: all those in forest streams died within a few weeks whereas a high proportion survived in adjacent moor-land streams. Stoner, Gee & Wade (1984), in a similar experiment with translocated young trout in the upper

A small stream completely overgrown by thicket forest. While planting is now held back from the banks of larger streams, it is scarcely possible to avoid enclos-ing the smaller streamlets and rills. Water chemistry and aquatic organisms are affected accordingly. Polmaddy Glen, Carsphairn, Galloway, 1985.

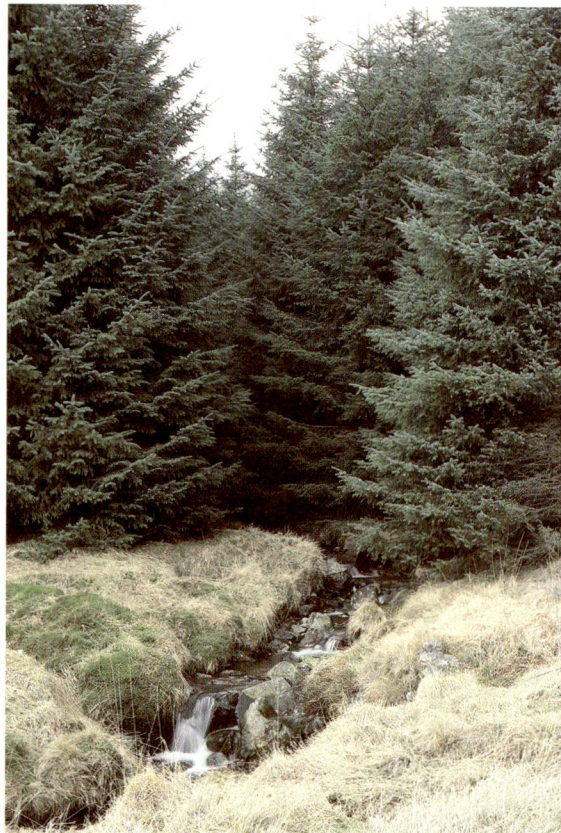

Environmental changes caused by afforestation

Tywi, found that survival after 17 days was 0-40% in afforested streams and 60-100% in moorland streams. Both groups of workers attributed the mortality effects to high acidity of the forest streams combined with the toxicity of the associated elevated soluble aluminium levels. Stoner & Gee (1985) regard the failure of attempts to stock the Rivers Camddwr and Tywi above the Llyn Brianne reservoir with salmon and sea trout as the result of elevated acidity and dissolved aluminium after extensive afforestation of the catchments. They also found that, in three lakes largely surrounded or bordered by recent conifer forests, brown trout catches declined from reasonable levels in 1971 to very low levels or zero in 1982, and that attempts at restocking failed, whereas reasonable catches were maintained in two unafforested lakes in the same area. The same picture of lower pH and higher aluminium levels in the afforested lakes was again found.

Mills (1980b) and Smith (1980) found that when forests encroach too closely on stream banks, the shading effect and intense needle-fall can reduce photosynthesis and productivity of aquatic plants, and hence the populations and variety of dependent invertebrates and the performance of both salmon and trout. With other workers (Stewart 1963; Graesser 1979), they also regard the combination of increased scouring and sedimentation (especially silt) which occurs in stream courses through ploughing for afforestation as especially harmful to the spawning beds of salmonid fish. Existing deposits suitable for the fish are scoured out or buried by fresh material. Increased suspended sediment and more variable flow regimes, with decreased low water flows, are also regarded as unfavourable to fish. By contrast, increased levels of nutrients such as calcium and phosphorus from fertilisers are likely to produce greater productivity for fish, provided that pH does not remain too low. Egglishaw (1985) has reviewed the effects of afforestation on fisheries.

Decline in one upland aquatic bird has been attributed to increased stream acidity resulting from afforestation. Ormerod, Tyler & Lewis (1985) found that dippers had disappeared from the uppermost 8km of the River Irfon in South Wales, whereas they were plentiful in the mid 1950s before the planting of a large proportion of the catchment. This was also correlated with a substantial impoverishment of the macroinvertebrate fauna on which the dipper feeds. Common sandpipers and grey wagtails continue to breed along the larger streams when the forest edge is kept well back from their banks. No declines in fish-eating birds are reported from afforested areas, but colonies of black-headed and common gulls have disappeared from tarns and swamps which became enclosed within forest. Botanical changes are to be expected from increased water acidity but have not been studied. Under the contrasting circumstances of nutrient (especially phosphorus) enrichment, increased growth of aquatic macrophytes and algal blooms are reported locally in lakes and reservoirs fed by streams from afforested catchments (Gibson 1976; Harriman 1978).

Pesticide use in forests

One of the features integral to the modern approach to afforestation is its recourse to the synthetic organic pesticides of agriculture. In Britain these chemicals are used in forestry only on a minor scale compared with agriculture, and foresters have shown concern to minimise both their use and any potential environmental consequences. For instance, the cumulative area of FC land sprayed with insecticide over the last 50 years is only 2% of the plantation area (Forestry Commission data). Aerial spraying is also limited to a few serious pest species. Pesticide applications in forests have nevertheless been quite widespread, and afforestation has had the effect of introducing these toxic chemicals to areas and habitats which had previously been almost pesticide-free. Lindane (HCH) is the only organochlorine insecticide still recommended for forest use, and nearly all the pesticides employed nowadays have low persistence and low toxicity to mammals, birds and fish (Forestry

Environmental changes caused by afforestation

Commission 1983a, 1983b). All these chemicals have been cleared by the Government's Pesticides Safety Precautions Scheme, on which NCC is represented.

The most serious forest pests are insects, either species feeding on foliage, buds or roots, or those boring into bark or wood, including types carrying pathogenic organisms. These can reach such numbers as to cause unacceptable tree mortality, deformation or retardation of growth. Several native moths and other insects can cause periodic outbreaks of damage to various conifers, including the indigenous Scots pine: they include pine looper, pine shoot moth, winter moth, pine weevil, black pine beetles, pine sawfly and green spruce aphis. Several introduced insect pests have also appeared (see Stoakley 1986 for an account). Relatively recently a more serious problem has emerged through epidemic outbreaks on introduced lodgepole pine of the native pine beauty moth, which seldom causes significant damage to its normal host, the Scots pine. Whole plantations of lodgepole pine on blanket bog in northern Scotland have been wiped out by pine beauty infestations. Population behaviour of the moth is therefore monitored and precautionary aerial spraying programmes are undertaken when epidemics appear imminent. Normally, the organophosphorus insecticide Fenitrothion is used, and since 1977 at least 15,000 ha have been treated, including 4,700 ha in 1985,though some areas have been sprayed twice. This appears to be a chronic problem in certain districts, and further control measures in 1986 are indicated (Watt 1986). Experimental monitoring of spraying over limited areas has not shown harmful population effects on any vertebrates (Crick 1986). Hamilton & Ruthven (1981) found, however, that there appeared to be a significant depression in brain acetyl-cholinesterase activity in sampled chaffinches and coal tits, giving possible occurrence of sub-lethal effects. Insect numbers generally are inevitably depressed within the treated areas, and their capacity for recovery is not yet known. The possible effects of spray drift outside the forest have not been examined. While the ultra-low volume technique of application (i.e. fine mist) is an efficient method of application,the risks of spray drift in northern Scotland must be high, even though critical limits to wind velocity have to be observed in spraying operations. Spray programmes have also been variably used in different areas against pine looper and pine sawfly.

Concern about pesticide use in new British forests arises partly from observation of the 'treadmill' effect of pesticide dependence in certain parts of North America, where repeated cycles of spraying have become routine in forest management. There is a widely held view, supported by the lodgepole pine/pine beauty moth example, that the growing of mono-culture crops, especially of introduced species, is a classic formula for increasing the risk of epidemic pest outbreaks. The experience of agriculture supports this, though part of the agricultural problem appears to be in the genetic uniformity of highly bred strains of the crop species, which does not at present apply to forest crops. There is inevitable anxiety about the recent appearance in Britain of the great spruce bark beetle, in view of the immense area of Sitka spruce already planted. This species, which can cause quite serious tree damage, has become widespread in the Welsh Borders. It is not susceptible to control by insecticides, and FC has introduced a beetle predator *Rhizophagus grandis* in an attempt to achieve biological control. Another non-chemical means of pest control regarded as showing great promise is the development of viruses which attack the target species. Progress has had to be cautious because of possible human health hazards, but the engineering of highly specific viral agents holds much potential for the control of particular insect pests. Good results have already been obtained in virus control of pine sawfly over limited areas. Biological control is always preferable in principle to the application of broad-spectrum biocides, although the introduction of

42

Environmental
changes caused by
afforestation

non-native predators is itself attended by an element of risk. Conservation ecologists will, however, always be concerned that attention be given to the development of silvicultural methods which reduce the risk of pest epidemics.

Chemical control is customary for those tree diseases of some importance in nurseries or plantations (Forestry Commission 1983a). The butt-rot fungus *Fomes annosus*, which develops in cut tree stumps and infects restocked growth, has generally been controlled by applying urea to the stumps, but another fungus, *Peniphora gigantea*, has been used in biological control in pure pine stands. Herbicides are used on better-quality ground to suppress other plants which compete with the young trees, slowing their growth or preventing their establishment (Forestry Commission 1983b). Planting in upturned ridges, turves or scarified patches in the uplands gives the trees a weed-free start, and weed problems are mostly on lowland sites: species sometimes needing control are grasses,

broadleaved herbs, bracken, heather, brambles, gorse, broom and other shrubs including broadleaved tree saplings. Herbicides are applied only when strictly necessary to benefit the trees, mainly by spot applications, and every effort is made to avoid contamination of watercourses. Nevertheless, FC received clearance for aerial application of glyphosate over up to 444 ha in 1985 (Hansard 1985b). This was either for grass control over newly or recently planted areas or to control self-regenerating birch on restocked sites.

The amelioration of physical and chemical effects of afforestation

Both the Forestry Commission and the non-state sector have become much concerned to develop methods which avoid or counter the adverse effects on soil and water associated with afforestation. Research on these environmental problems has been expanded and management studies commissioned, for example *The management of forest streams* (Mills 1980a), published by FC and especially relevant to the conservation of fish. It is now standard practice to

Plantation of lodgepole pine completely killed by infestation of pine beauty moth. This insect is now controlled by spraying with the organophosphorus insecticide Fenitrothion. Rimsdale Forest, Badanloch, east Sutherland, 1980.

Environmental changes caused by afforestation

keep the forest margin well back from the edge of the larger streams (a distance of ten times stream width, up to 30m), thereby preventing shading and litter fall and reducing disturbance and acidification effects to the water. Ploughing technique has become much modified, and drains are stopped 15-20m short of natural watercourses, so that the water has to filter through soil or ground vegetation before reaching the stream. This has beneficial effects in reducing rate of run-off, amount of transported sediment and nutrient loss, and in neutralising water acidity: the last effect can be enhanced by ploughing more deeply into the subsoil. The most acceptable course of all is the increased replacement of ploughing by 'ripping' or spot scarification, which reduces disturbance to hydrology. Roberts (1985) has suggested further ways in which nutrient losses at harvesting stage can be reduced. Fertiliser applications now aim to minimise losses through run-off and are no longer permitted under extreme weather conditions, whilst pesticide

usage has steadily sought to eliminate environmental hazards.

The general application of these beneficial modifications is greatly to be welcomed, and the search for still further improvements encouraged. It is, nevertheless, necessary to point out that in most respects they represent an amelioration and not a cure for the problems. Moreover, the measure of their actual success will not become clear unless or until adequate monitoring programmes provide hard data. Maintaining buffer strips to larger streams is undoubtedly beneficial, but there is little possibility of afforestation avoiding disturbance to most of the innumerable springs, rills, flushes and seepage areas which are integral to any upland drainage system but are scattered through the planted ground. It is also inescapable that on badly drained ground, varying from wet gley soils to deep raised or blanket bog peat, a main purpose of ploughing is to lower the water table and dry out the ground surface so that the trees can grow; in other words, major hydro-

Banks of larger stream kept clear of forest. The now little-grazed vegetation has become more luxuriant and is good invertebrate habitat. The young forest of Sitka spruce is closing to the thicket stage, excluding other vegetation. Tarras Water, Langholm Hills, Dumfriesshire, 1980.

4

**Nature
conservation
and afforestation**

**Environmental
changes caused by
afforestation**

logical disturbance is the actual
objective. On these wet sites, it is difficult to conceive how ploughing can be
avoided. And as long as there is
ploughing there will be increase in both
run-off and erosion. Material is bound to
erode from the sides of drains, though it
may remain trapped when the ends of
these are blocked. Newson (1985) notes
that stopping of ploughing and ditching
well short of stream courses has little
effect on fine silt load even if it controls
gravel yields. Even the technique of
'ripping' is said to cause a surprising
amount of soil erosion.

   Improvements thus seem likely to
be most marked on dry sites, and least
on the extensive blanket bogs where so
much new afforestation is now occurring. The adverse effects of previous
practices will remain, but it is to be
hoped that their significance will
diminish in the longer term. Eventual
reduction in acid deposition at source is
likely, and in some areas where the
overall acidification levels are high,
there are attempts to counteract the
trend, notably by the experimental
liming of lakes and sometimes their
inflow streams. Apart from the costs of
this operation, which has to be repeated
at intervals, there is the problem that,
far from restoring previous chemical
conditions, it creates a somewhat
different state which may favour certain
fish but does not bring back the
previous natural biological community.
Raising pH to previous levels by liming
causes an increase in calcium content
of water to levels higher than those
occurring previously, and probably
also an increase in toxicity of aluminium
(Fry & Cooke 1984). Nor can liming
cope with heavy rain and flood
conditions, which produce pulses of
high acidity and elevated aluminium
levels regarded as especially damaging to some organisms including fish.
The greater use of broadleaved tree
species in afforestation programmes is
beneficial to soil and water conditions,
and environmental conservation in
general would benefit from an increase
in their use in new planting and
restocking programmes.

Significance of afforestation changes to nature conservation

The nature conservation value of new forest and its enhancement

The main ecological consequence of the changes described is that afforestation causes an extensive replacement of open ground habitats, communities and species by those of forest. Foresters believe that this should be looked at positively, in terms of gains and how these may be enhanced in future (Steele & Balfour 1980). Descriptions have been given of how the new forests will evolve through subsequent rotations to a more varied and interesting state, losing their present uniformity as the opportunities for diversification of tree composition and structure are increasingly realised (e.g. Forestry Commission 1980; Hibberd 1985). Some mature forests managed with sympathy for wildlife are regarded as models which show how this potential can be achieved. Yet present potential cannot be relied upon to become future reality without further reassurance on how afforestation will develop and be managed. The new forest estate will not reach fruition until well into the next century, but conservation concerns need to be dealt with now.

A brief description has been given of the cyclical wildlife succession in a new conifer forest (see pp. 26-33). The interest is especially in the birds and mammals, though the invertebrate fauna can be quite rich in the areas of open ground. Botanical features tend to be the least interesting. Wildlife variety is usually least in forests with large, even-aged, minimally brashed and thinned stands of single tree species, with lowest representation of open habitats. Management for wildlife interest should thus aim in the opposite direction as far as possible. The particular measures which favour plant and animal variety are already well known and have been dealt with by Steele (1972).

Varying the choice of exotic conifer species has relatively little effect, but wildlife value increases with the number and proportion of native tree species, mainly broadleaved though including Scots pine. Native tall shrubs are also valuable. The best effect is achieved by planting or encouraging these in clumps and belts, especially along the margins of rides, roads or other open ground. Birch, rowan and willows are the most readily self-regenerating species: others usually have to be planted. Where there are existing native trees, either scattered or especially as blocks of woodland, it is important that they be retained and suitably integrated with the new forest.

Designing the forest to retain a reasonable amount of open ground is important, especially on sites where all the land is plantable. Rides should be wide and areas of grassland, dwarf shrub heath, bogland and scrub should be left unplanted, or planted only with an open scatter of native species. Such areas should not be ploughed, and it is important to keep patches of wet ground. The creation of small areas of standing water is extremely beneficial. Unplantable habitats such as rocky ground, spongy bogs and existing open water are valuable for wildlife, as is the more recent practice of keeping plantations well back from major stream banks.

The conifer stands will nevertheless continue to occupy most of the forest area, and the management of these is most important. Unless they can be well-thinned at appropriate stages, the possibilities for enhancement of wildlife interest are strictly limited. The more that forest can be allowed to reach full maturity as well-thinned stands, the greater will be its botanical and faunal variety. Dead trees, both standing and fallen, and especially the larger individuals, are a most necessary habitat and should be left to decay. It is also valuable to allow some trees to become over-mature and even to die of old age, as in a natural forest. For good effect, there should be as much well-thinned, mature forest as possible at any one time.

A major opportunity for diversification comes at felling. Large areas of uniform age-class can be felled over a period of years, in blocks of different shapes and sizes, to give a greater spread of age-class diversity in the next generation forest. This spread of harvesting is facilitated by longer rotations but is reduced in districts

5

**Nature
conservation
and afforestation**

**Significance of
afforestation changes
to nature
conservation**

where high windthrow hazards lead to early felling. Opportunity to diversify continues through subsequent rotations, so that, in theory, it should be possible to end up with a range of age-class diversity from clear-fell to maturity. The open ground phase (pre-thicket) in each rotation is important and can be increased by deferring replanting: the larger the proportion within any forest at any one time, the better the opportunities for clearance-phase wildlife. The tendency to reduce the size of any one area (through staggered felling) will militate against the return of open ground birds needing large nesting territories, so that felling of blocks adjoining recently cleared ground may sometimes be better than distributing the clearance phases through the forest.

Fine-scale structural diversity, in which age of trees varies widely and irregularly within a single stand, is the most difficult feature to achieve in a commercial forest. When enough saplings are present in a maturing stand, they can form a partial medium to tall shrub layer — the feature most constantly absent in conifer plantations. A shrub layer is a valuable animal habitat present in many ancient semi-natural woodlands where it is composed of species such as juniper, hazel, blackthorn, willow, holly and rowan.

More specific habitat management, outside silvicultural modifications, may be necessary for some wildlife. Many birds and mammals will colonise suit-able forest areas of their own accord, but many invertebrates and plants will occur only if they are deliberately put there. Many species have such limited capacity for increase and spread that even then their abundance will depend on numbers introduced. Some species, for example bats and hole-nesting birds (from titmice to owls, kestrels, sawbill ducks and goldeneye) can be encour-aged by providing cavity roosts and nestboxes. Rare species may need quite specific measures including habitat manipulation and careful protection; examples include military orchid in Breckland, sand lizard in Dorset and goshawks more widely.

Each forest needs to have its opportunities assessed and appro-priate prescriptions made. Many modifications to forest management for reasons of scenic amenity and visual improvement are also beneficial to wildlife.

The possibilities for geological enhancement seem to be limited to the exposure of new and important sections or materials through excavation work, whether in quarrying road material or making cuttings in the routing of the roads. In some localities it may be possible to use geological knowledge to increase the value of such exposures.

Natural ecological diversity, such as range of topography and altitude, variation in soil types from acidic to calcareous, and past differences in land-use, can add greatly to the wildlife value of any site. Any features which prevent tree planting, such as spongy swamps and open waters, rock habitats and ground above the climatic tree limit, will give an intrinsic element of diversity. This kind of variety is, however, a fortunate chance of nature and should be distinguished from that which can be planned and created. On many afforested areas, virtually all the land is plantable, apart from minimal requirements for rides, roads and other necessary open ground. Any other ecological diversity remaining after afforestation will depend entirely on specific planning to accommodate wildlife interest. The general rule is that the more that can be achieved for any one of the measures listed above, the better the results for wildlife, and that their value is additive. Some loss of timber production has, however, to be accepted from these practices.

Within the limitations of dependence on exotic conifers, the opportunities for diversification of the new forests to increase their value for wildlife and to give widely ranging ecological conditions are very large. But how far will these opportunities be realised in practice? This depends on how far timber production, and hence profit margin, will be forgone in the interests of wildlife. Established forests produce a larger income as they grow older, so that costs incurred in the

47

Significance of afforestation changes to nature conservation

interests of wildlife may become less burdensome. Yet these costs will remain; so statements that the new forests will become far more interesting for wildlife than at present must reflect a willingness to incur proportionately greater financial penalties in future for the benefit of wildlife.

Another crucial factor is the strict environmental limitation to implementing some of the desirable management options. High windthrow hazards have led locally to a preponderance of thicket-stage plantations and early harvesting. A large proportion of the new forests established in the last 20 years have been in districts and on soils with high windthrow hazard. Improvements in silvicultural and management techniques may lead to some amelioration of the windthrow problem, but it is unlikely that we can obtain the well-thinned, mature stands of the kind mentioned on page 46 in the areas concerned. Much of the new planting is on poor soils and peats where the scope for diversification of species is very limited.

Prospects for obtaining improvements for wildlife increase as environmental conditions become more benign. Some of the best examples of new forest with relatively high value for wildlife are in Breckland (notably the King's Forest), where there have been: retention of blocks of existing native woodland; creation of 'amenity fringes' of broadleaves, wide rides and open areas; and fairly heavy thinning of many compartments, allowing extensive development of field layers. Grizedale Forest in southern Lakeland and Coed-y-Brenin in North Wales are regarded as other demonstration areas with emphasis on diversity in relation to wildlife management. Plantations on the North York Moors and around Betws-y-Coed in Snowdonia also show good integration with existing broadleaved woodland and hill farming. At Eskdalemuir in the Southern Uplands of Scotland, special attention has been given to deer and other game management in forest design. Wildlife interest of the new forest tends to increase in a southerly and easterly

Amenity fringe of birchwood to pine plantations. The planting of broadleaved trees or incorporation of existing stands within new forests is an extremely beneficial form of management for wildlife interest. King's Forest, Suffolk, 1980.

Significance of afforestation changes to nature conservation

direction and with decreasing elevation: for, as climate becomes more favourable and soils more fertile, the opportunities for beneficial management increase.

Provision for wildlife in forest design and management has so far been extremely variable. There are forests which reflect sensitivity in these respects, and these show how some of the opportunities have been realised. They are much to be welcomed and encouraged in future. But there are others where the wildlife content owes virtually nothing to choice and everything to circumstance, and the greater part of the new forests, especially in Scotland, shows few if any concessions to nature conservation. Such improvements as have been made are usually for scenic rather than nature conservation reasons.

The wildlife opportunities in the new forests have to be approached realistically, with willingness to talk about the problems and to acknowledge the limitations. NCC is anxious to help, by advising on whole forest plans as well as those for special woodland sites, and has recently responded to a Forestry Commission invitation to contribute to revision of the plans for its Breckland forests. While there is much scope for further research into forest management to benefit wildlife, especially on the precise requirements of individual species, NCC sees the main needs as field experimentation, with combinations and applications of known measures, and the monitoring of the results. However, support and co-operation to improve forest management for wildlife does not reduce our concern over the negative side of afforestation, namely the losses caused to the resource of wild nature on previously open ground.

The losses to open ground wildlife and physical features

Afforestation causes a replacement of open ground habitats, communities and species by those of forest. However varied it may become and whatever the modifications to afforestation practice, the new ecosystem will remain essentially forest, and therefore quite

A harmonious mixture of hill farmland, native broadleaved woodland and coniferous plantation, illustrating integrated land use. Many lower income farmers find it difficult to engage in new planting, other than small shelter woods, because present financial incentives are weighted to tax relief instead of towards direct grant aid. Near Llangollen, Denbighshire, 1984.

Significance of afforestation changes to nature conservation

different from the type it replaces. The one will increase and the other will decline. The nature conservation concern over this fundamental problem can be best explained by addressing certain questions that are repeatedly put by foresters.

A reasonable representation of open ground habitats and wildlife survives within the forest areas, and further provision can be made for this, so are not the changes over-stated?

Survival of open ground wildlife varies greatly between forests. At worst it is a tiny and insignificant remnant and at best an incomplete and inadequate representation. The combination of wildlife of open ground and scrub that is characteristic of the first ten years of a new forest is a 'once-off' condition, for the pre-thicket phases of subsequent rotations will have lost substantially more of the open ground features. Shortening the rotation increases the proportional area of pre-thicket stage, but this does little to restore the lost habitats and wildlife. The linear open habitats of rides, road verges and stream edges have value for some wild-life, such as butterflies, reptiles and amphibians, some mammals and a few riparian birds, but they are too narrow to allow the nesting of most birds of open ground. And while these habitats support some remnants of previous vegetation, their ribbon or net-like form is an artificial representation, and it usually precludes survival of unmodified wet ground systems of springs, flushes and bogs which require drainage to remain undisturbed over much larger areas. Afforestation has a particularly destructive effect on these wet ground habitats.

A list of open ground species can be compiled within the treeless habitats of some forests, but some of the most valued species do not survive or return and many of the others do so only in insignificant numbers. The nature conservation value of any ecosystem is not to be measured just as a species inventory. It lies in the communities, both plant and animal, in the distribution and abundance of the species themselves, in the varied physical attributes of the habitats, and in the relationships of all these features to each other as an intact and complete biotope. On many areas where nearly all the ground is plantable, afforestation causes an ecological transformation. As an ecosystem, whatever was there before — a blanket bog, for instance — is virtually obliterated, and the forest which develops represents a complete contrast.

Other open areas remain in some forests, usually because they are unplantable or because the trees have failed and not been replaced. Rock habitats have their own specialised flora and fauna, some of which may benefit from the protection of the surrounding forest, provided that this does not encroach and give too much shade and litter fall. Patches of especially spongy bog, pool systems or dry ground habitats sometimes remain as permanent open areas and are welcome as such, but they are usually too small to have great value. They may represent samples of vegetation, but divorced from their wider ecological context and usually amounting to rather unsatisfactory relics. Their area is usually too small to support a characteristic moorland bird community, but they may retain a fair invertebrate fauna. In some districts, larger areas of hill top are left unplanted above the forests. The wild-life value of these depends very much on their area and altitudinal range, and on whether they remain grazed. The smaller areas tend to have a depleted bird fauna, but for some animals, especially invertebrates, they can be important. The vegetation of ungrazed and unburned enclaves shows the same kinds of change that occur within newly planted ground, with dwarf shrubs extending their cover but no new species appearing. On the higher Scottish mountains, a broad zone of important montane habitats lies above the forest zone. Yet only a limited fauna lives wholly within this montane zone, and the greater part of the animal life in the uplands depends essentially on the more productive submontane afforest-able ground below. Some larger species, such as golden eagles and ravens, which feed and even breed on

Significance of afforestation changes to nature conservation

the higher ground, can have their food supply reduced to critically low levels if forest occupies too much of the lower slopes. Red deer find much increased shelter within the tree stands, but often have to feed mainly on open habitats within the forest or beyond.

Forest and open ground are different, but surely both types of ecosystem are desirable, whether within a single open ground site or a larger area?

The argument that a limited amount of afforestation will add further diversity to the best open ground sites has already been refuted (see p.13) Orthodox afforestation has a highly destructive effect on the perception of wildness and on attempts to maintain or enhance naturalness, because the diversity it creates is conspicuously artificial. It can also cause an unacceptable loss of valued habitats and increased predation risks to breeding birds (see p.33). Such developments cause greater losses than gains to nature conservation interest on sites which are especially valued for their open ground features,

and they are disruptive and damaging to such interest.

Viewed on a larger geographical scale, a reasonable proportion of new forest in some districts can be a benefit to wildlife, as a woodland habitat. Much depends on where the plantations are located and on their extent. From a nature conservation viewpoint, afforestation should concentrate on ground of lowest importance for wildlife and physical features, and its total extent should not exceed a reasonable balance with open ground in any district. Enhancement of diversity can become a useful feature on this larger scale, but this does not necessarily support the argument for having some afforestation in all districts, as advocated, for example, by Helliwell (1978).

Mosaic planting has been regarded as a more desirable option than blanket afforestation; it can give good opportunities for integrated land-use, as well as a combination of open ground and forest habitats (e.g. Newton 1984). In some parts of central Wales,

Blanket afforestation with conifers. On low moorland little ground remains unplanted when it is acquired by forestry interests. Roads and rides provide linear open ground habitats, but contain only a fragmentary representation of the previous moorland ecosystem. Kielder Forest, Northumberland, 1985.

Significance of afforestation changes to nature conservation

this patchwork of forest and hill farm has succeeded in maintaining the good populations of ravens, buzzards and red kites. But in areas where the ornithological interest is especially for waders and grouse, the problem of increased nest predation again arises, and the mosaic may need especially careful planning. However, mosaic planting must not be allowed to become the stepping stone to blanket treatment. The planting of plateau bog at Llanbrynmair Moors, Powys, was given as an example of successful integration between forestry and farming: since then, several more farms within the area have been sold to forestry. This underlines the importance of maintaining the fabric of farming communities; farms which become isolated in an afforested area can lose viability and as a result may themselves be sold for afforestation.

Surely forests are more interesting for wildlife than open ground, and will become increasingly so in the future? They are no more artificial than most open ground habitats and are, indeed, **closer to what was there originally.**

Undeveloped land varies widely in its interest for wildlife and physical features, and this variation is implicit both in the concept of special scientific interest and in the fuller set of values which has come to determine the practice and priorities of nature conservation. According to the criteria that have consistently been used to assess biological aspects of nature conservation value in Britain, exotic conifer plantations rate as less important than areas judged to be of special scientific interest (see p.13).

One simple index of nature conservation value is 'non-re-creatability' (see p.15). Many of the most important wildlife habitats and physical features are, for all practical purposes, irreplaceable once they have been lost. Some (such as Breckland grass-heaths) might be restored at considerable difficulty and cost, but there is always an especial problem in replenishing the full species complement, and the areas would have to be large enough to be significant.

Aerial view of afforestation in the flow country. The patch of patterned pool and hummock bog will remain as an island within the forest, but its hydrology may become altered and its relationship with the rest of the moorland ecosystem is lost. Many of the moorland birds need much larger open areas than this.
East Sutherland, 1985.

Significance of afforestation changes to nature conservation

Once the abiotic environment is significantly modified, restoration is usually still more difficult and change tends to be permanent. There is no possibility of re-creating most peatland habitats on a significant scale or within a time-scale we can contemplate — even in the unlikely event that the land will later be released from forestry or other uses. Effectively, habitats lost to afforestation have gone for good, and they cannot be redeveloped elsewhere. By contrast, new forest can be created on a wide range of sites, and its quality either commercially or as wildlife habitat usually has little if anything to do with the nature conservation value of what it replaces. New forest can be greatly improved for wildlife by careful management, but this does not validate planting on the best wildlife sites.

The new forests **are** more artificial than the habitats they replace, because they fall below the definition of semi-natural. While the new forests bear a superficial resemblance to those of the Boreal coniferous forest zone, a comparison with such forests in Fennoscandia or Canada, or even with the native pinewood remnants in the Highlands, shows that the similarity is spurious. Most of the natural Boreal forests are characterised by a continuous and dense ground vegetation of dwarf shrubs, herbs, ferns and their allies, mosses and lichens, whereas ground vegetation is largely absent from the thicket plantations which appear to be increasingly the norm in Britain. Many Boreal forest regions also typically contain substantial areas of open ground, especially completely unmodified peat bogs of all shapes and sizes, showing natural transitions to the surrounding forest. Natural processes, such as forest fires, followed by regeneration, produce a more random distribution of different age-classes, and extensive areas of open tree growth, classified as pine heath or spruce heath, are equally characteristic. Some of the Boreal forests, notably in Alaska, Canada and Russia, also have a complete and natural predator force including, importantly, the major predators of deer, which are absent from all British forests. Moreover, the greater part of the new British forests has been established in areas and situations where the original forest was not Boreal coniferous, but cool temperate broadleaved, and comparisons should therefore be with the latter. Forests are sometimes located in areas where the ground is too wet for trees to grow naturally. The fauna contains few of the Boreal forest species, and when the modifications to the physical and chemical environment are also considered, the claim that the new forests are restoring earlier forest conditions is not tenable.

Many of the characteristic wildlife species of the new forests are widespread plants and animals of various lowland habitats which do not represent a conservation problem, whereas many of the open ground species are restricted to these habitats and/or declining generally. The most typical birds of new forest are garden songbirds, but only the meadow pipit and skylark amongst the open ground species could be regarded as equally common. Waders such as snipe, redshank and lapwing have so declined in the lowlands through farming improvements that the uplands and marginal lands have become their chief strongholds. The international scarcity of many of the characteristic plant communities widespread in Britain, and the wide variety of these, have already been stressed. The new forests are much less varied, especially within the tree compartments, and most of their variety derives from the residual elements of open ground habitat. Blocks of Sitka spruce have a considerable uniformity across Britain and most of their variation is in the animal communities. The general tendency to drying of wet ground under afforestation reduces the scope for variety within both restocked areas and remaining open habitats.

New forests, of whatever type, can only strictly be compared with other forest types: they are fundamentally different from the ecosystems they replace and so cannot be compared with them. The question posed and the argument used to support it are thus really about preference and choice,

Significance of afforestation changes to nature conservation

Typical maturing stand of Norway spruce, about 40 years old, brashed and moderately well thinned. Ground cover a mixture of litter with sparse grasses and moss patches: shrubs absent. Songbird habitat. Cairn Edward Forest, Loch Ken, Galloway, 1979.

Native Boreal conifer forest of mature Scots pine with understorey of juniper and field layer of bilberry, cowberry, heather and moss carpets. Capercaillie habitat.
RSPB Loch Garten Reserve, Abernethy Forest, Speyside, Inverness-shire, 1976.

with an accompanying rationalisation. Some people prefer woodlands and their wildlife, and it can be shown that species variety and populations of birds are sometimes larger than on the previous open ground, and so on; but to argue that this makes the new forests 'better' than the replaced habitats is to misunderstand the principles of nature conservation evaluation (see pp.12-16). The issue is often clouded by pejorative and prejudiced references to open ground habitats with their existing land-uses as 'deserts', 'valueless and useless', 'sterilisation of land' and so on. Both habitats have value in their own right, and whatever the economic or other arguments for planting open ground of lowest wildlife interest, there is still a strong case for retaining a reasonable extent of open ground in any one district.

There may have been some losses, but surely there is and will always be plenty of open ground, covering the range of variation, left elsewhere?

This issue has to be seen in a historical context to appreciate the nature conservation perspective: it can be likened to a moving film which runs on, though we have chosen to arrest it momentarily at an arbitrary point in time. Since 1919, well over one million hectares of new forest have been created, covering over 4% of Britain. The planting has, however, taken place mainly in the hills and moorlands, which in 1919 covered about seven million hectares, including a proportional area of the existing woodlands (say, 300,000 ha) and also 2·5 Mha of unsuitable ground (Table 1). The national proportion of this plantable submontane terrain which has since been afforested is thus around 27%. The rate of expansion is perhaps better conveyed by the fact that over the last ten years, 1976-85, new afforestation has covered 226,800 ha, an area almost the size of Cheshire. It has, moreover, shown a distinct ecological/geographical bias, falling most heavily within the major remaining expanse of semi-natural habitat in Britain.

Past reliance on the market availability of land for afforestation (see p.75) has resulted in a quite substantial amount of loss and damage to areas important for wildlife and physical features. Some highly localised types have lost ground. The unique mixtures of calcareous grassland and acidic heath in the Breckland have been reduced by about 19,500 ha and significant proportions of remaining lowland acidic heath in east Suffolk and Dorset have also been planted. Large sections of the important sand-dune systems at Newborough Warren, Anglesey; Ainsdale, Lancashire; Torrs Warren, Wigtownshire; Tentsmuir, Fifeshire; Culbin Sands, Moray and Nairnshire; and Dunnet Links, Caithness, have been converted to forest. Raised bog is an extremely localised type of peatland, and afforested areas include Foulshaw Moss, Cumbria; much of the Lochar Moss complex, Dumfriesshire; Auchencairn Moss, Kirkcudbrightshire; much of Moss of Cree, Wigtownshire; and much of the West Flanders Moss system in Stirlingshire. Calcareous and other base-rich habitats are not often planted, but botanically rich pastures and heaths at Greystoke Park and Johnby Moor, Cumbria, have been planted. Planting of conifers on the basaltic crags and screes at Stanner Rocks, Radnorshire, and on the dissected limestone pavements at Whitbarrow Scar, Cumbria, has reduced the naturalness of these very localised calcareous habitats.

While much of the afforested ground was rated as average to low in nature conservation quality, some areas representing especially good examples of widespread habitats, such as dwarf shrub heath and undamaged blanket bog, have also been lost. They include Llanbrynmair Moors, Powys; part of Denbigh Moors, Clwyd; much of the North Tyne and Bewcastle Moors, Northumberland and Cumbria; Kilquhockadale Flow, Mindork Moss and Annabaglish Moss, Wigtownshire; Arran Northern Mountains, Buteshire; part of Kerloch Moor, Kincardineshire; and several large and important areas of the east Sutherland-Caithness flows.

Afforestation has become, in terms of extent, the principal agent in Britain of the replacement of semi-natural

Significance of afforestation changes to nature conservation

habitats by those dominated by introduced species, and this geographical effect could become doubled eventually. There is little information on the extent of the widespread plant communities converted to new forests. In variety, they include virtually all the types listed in *A nature conservation review* (Ratcliffe ed. 1977), Volume 1, on pp. 293-297 under submontane dwarf-shrub heaths, acidic grasslands, fern communities, wet heath, blanket mire, soligenous mire, and spring and flush vegetation for the uplands; on pp. 252-253 as raised mires; on pp. 134-136 as lowland heaths and grasslands; and on pp. 29-30 in respect of stable dune heaths. Within the more refined National Vegetation Classification (in preparation), at least 107 out of 261 (41%) of recognised plant communities have been affected by afforestation somewhere in Britain (J Rodwell, unpublished). Blanket bog and heather moorland will be increasingly depleted in future, and at present the only submontane habitats where afforestation is constrained are on the areas of fertile soil, on limestone or other basic rocks, which the Agriculture Departments have in the past been reluctant to release.

Populations of open ground species have been variably affected by afforestation. While no plant or animal species has become extinct in Britain or even seriously threatened with extinction through planting, some have declined appreciably. The characteristic plants of the communities lost to conifers have declined in proportion, though most still remain widespread and abundant. Several rare or local plants have, however, lost a significant number of their stations or suffered substantial population depletion within certain counties. This is especially true of the distinctive flora of Breckland, although, in the absence of afforestation, agriculture might have caused an even bigger depletion. Some notable plants of lowland acidic heath have lost ground, for example petty whin, dwarf gorse, lesser dodder, marsh clubmoss, stag's-horn clubmoss, marsh gentian, the sundews, the butterworts and

Native Boreal conifer forest of Scots pine, varying in age from seedlings to 200+ year old trees. There is good habitat diversity, with widely spaced trees and dense field layer, open heathery and boggy clearings, and dead pines which provide habitats for insects. Glen Tanar Forest, Deeside, Aberdeenshire, 1964.

Significance of afforestation changes to nature conservation

Dorset heath. On blanket bogs and raised bogs, bog rosemary, cranberry, dwarf birch, mud sedge, tall bog-sedge, few-flowered sedge, white beak-sedge, black bog-rush, the bladderworts, and rare bog-mosses such as *Sphagnum imbricatum, S. fuscum, S. pulchrum* and *S. balticum* have declined; and in wet base-rich habitats so have grass-of-Parnassus, globeflower, common spotted-orchid, marsh-orchids, broad-leaved cottongrass, long-stalked yellow-sedge, lesser clubmoss, and the rare moss *Camptothecium nitens.* Few uncommon plants have benefited from afforestation, and the dune helleborine on Newborough Warren and Ainsdale Sand Dunes is perhaps the best example. Decline in bird populations is mostly difficult to measure, but the raven had shown a reduction from at least 100 pairs in 1955 to only 35 pairs in the Southern Uplands and Cheviots by 1981 (Mearns 1983), and 85% of the loss is attributable to loss of foraging ground through afforestation. Widespread moorland breeders such as curlew, golden plover, dunlin and red grouse have all suffered appreciable population declines through afforestation in Wales, the Cheviots, the Southern Uplands, Sutherland and Caithness; and the losses to their numbers in Britain could eventually become substantial (NCC data). From known average densities, it is estimated that if all the 67,000 ha of Sutherland and Caithness presently owned by forestry interests were planted, about 250 pairs of greenshanks would be lost, and this is estimated to be 25% of the British population of this highly localised wader (Thompson, Thompson & Nethersole-Thompson 1986). Few effects on other rare animals are reported, but in northern Scotland one of two remaining mainland nesting sites for red-necked phalarope was recently destroyed (R. Dennis, unpublished), and on the heaths of southern England significant habitat loss for Dartford warbler, smooth snake and sand lizard has occurred, even though the last two species persist on open ground within forests (C. R. Tubbs, unpublished). It should be emphasised that the birds

Dubh lochans (black pools) with bogbean and cottongrass on the blanket bogs of the flow country. Habitat of a tundra type bird fauna, including greenshank, dunlin, golden plover, arctic skua, red-throated diver, common scoter and greylag goose. Afforestation has encroached within the area shown since 1980. Near Loch More, Thurso River, Caithness, 1980.

Significance of afforestation changes to nature conservation

and most other animals displaced by new forest do not simply move to other remaining open ground and become more concentrated there: they disappear, because these open areas usually already hold the maximum sustainable numbers of their kind.

The scale and rate of past losses to wildlife and habitat, especially to the more highly valued elements, compels us to look at the past and at the present for trends which give portents for the future. Such evidence for the continuing attrition of the remaining area of undeveloped, semi-natural habitat in Britain, through the whole range of human impacts, was presented by NCC and the non-governmental organisations to Parliament and led to appropriate strengthening of the wildlife legislation in 1981. The manner in which open ground habitats have been disappearing over the last few decades gives great anxiety that, before long, a great deal less will remain and that further depletion of the most important features will occur. In certain districts the total loss of open ground habitats is already so great that, for nature conservation, further losses are unacceptable. Yet it is not only the scale of loss that is so worrying, but also the way in which this has previously fallen indiscriminately on the more valuable habitats and areas as well as the less interesting ground. There is great uncertainty about how much open ground will eventually be left, outside the protected sites, in any district.

Nature conservation concern about changes to the physical and chemical environment

NCC's concern over the effects of afforestation on non-living aspects of the environment is mainly about their relevance for biological components, but relates also to its responsibility for conserving physical features. The Council's duty to take account of actual or possible ecological changes (see p. 7) was intended by Parliament to allow a broad view of the conservation implications of land-use practices and their impacts, in relation to sound environmental management. This is understood as including a responsi-

The wetter moorlands of the Scottish borders. Blanket bog with cotton grass, heather and cloud-berry is extensive on watersheds, with a mixture of dry heather and bilberry moor, and grassland. The haunt of golden plover, curlew, red grouse and merlin. Roan Fell, Tarras Water, Dumfriesshire, 1979.

Significance of afforestation changes to nature conservation

bility for taking the initiative in giving advice to Government and other parties on important ecological issues. Besides having a direct concern for environmental management on NNRs and SSSIs, NCC has a more general concern for the maintenance of quality of atmosphere, water and soil.

The degrees of hydrological disturbance, increased erosion and sedimentation, water acidification and nutrient enrichment (see p.35) seem to NCC at times to have transgressed the limits of sound environmental management and conservation practice. While appreciating that continual attempts at improvement in afforestation practice are sought and applied, NCC shares the anxiety expressed by interests with direct responsibility for these resources. Afforestation policies and practices must be questioned until much wider reduction in these disturbances has been achieved. NCC is concerned about conservation of aquatic life in general, including the rarer fish, and is deeply concerned over problems of

acid deposition. Fish are, moreover, good indicators of the 'health' of fresh waters. The issue of soil fertility within plantations is the concern of forestry so long as the land remains under trees, but it would be worrying if any long-term changes effectively ruled out any other form of use in the future. The cracking of deep peats under forest in northern Scotland should be monitored in this context. Adequate research input is a necessary part of the effort to overcome or reduce such problems.

Effects on geological and physiographic features

Tree planting has little direct effect on more massive geological features, but it is important to an appreciation of geological and geomorphological processes to be able to **see** features, and their essential character often becomes invisible under a blanket of trees. Even if certain small-scale features are left unplanted, their significance and importance are greatly reduced if they cannot be related visually to each other, which is the case once the plantations around

The patterned blanket bog of a National Nature Reserve, damaged at one end by draining for afforestation. The moorland ecosystem as a whole is becoming extensively dissected by new forest, and its character changed. Strathy Bog NNR, Sutherland, 1985.

Significance of afforestation changes to nature conservation

them have grown up. Both the visual appreciation and the research potential of large-scale geomorphological sites can be severely restricted by mature forestry plantations. The internationally famous Parallel Roads of Glen Roy, an altitudinal sequence of ancient glacial lake shores, are partly protected against afforestation as a result of an agreement between NCC and the Forestry Commission. However, other parts of the feature, which extend into neighbouring glens outside the Commission's land-holding, do not have such protection and are still threatened by planting proposals from time to time.

Especially in Scotland, the visual interpretation of extensive geological sites, such as ancient volcanic complexes, can be made difficult or even impossible by extensive planting. Outcrops of different rock types, whose spatial relationships have to be clearly visible on the ground if a geologist is to be able to interpret them, become invisible or unfindable. In the Glenelg area, nationally important exposures of Moine conglomerates resting upon Lewisian basement rocks are now impossible to reach because of afforestation. The softer land-forms and periglacial features are also vulnerable to real damage through afforestation. Moraine systems can be badly scarred by ploughing operations, as well as rendered invisible under a mantle of trees. Changes in erosion and sedimentation which occur as a result of the artificial run-off systems which arise from new planting produce unnatural features in river and lake systems and can also cause problems for cave conservation; increased rates of run-off change the hydrogeological character of cave systems and often have impacts on the cave sediments within them. Important sand-dune systems such as Culbin Sands, Newborough Warren and Ainsdale Sand Dunes have lost some of their geomorphological interest through partial planting. Old mineral working spoil-heaps which are of geological importance have been regraded and lost in a few areas. Planting close to the edge of important gorge and stream sections has caused some of these to become gradually

obscured by litter or moss and algal growth. Access roads are integral to the new forests, and softer land-forms such as morainic features are not only damaged by them but often form a source of road material. On the credit side, forest roads have cut new sections which in certain areas have proved to be of international importance (as in Mortimer Forest near Ludlow) and FC has agreed to maintain these sections for geological research and education.

Afforestation can reduce or destroy the value of sites important to Quaternary studies. Draining, surface drying and cracking of peat bogs can lead to disturbance to the upper peat and oxidation and loss of sub-fossil plant remains and can wash pollen down to lower horizons. Archaeological features of various kinds form an important link with biological prehistory, and their intercorrelations are crucial to an understanding of the ancient record of human development, and of change in biological and geological processes. Many archaeological remains lie close to the ground surface in the shallow soils of hill areas and so are destroyed by deep ploughing. While foresters have been concerned to protect important features drawn to their notice, the Council for British Archaeology is greatly concerned that afforestation has affected large areas which are rich in archaeological remains that have been little studied if at all. Here, there has inevitably been a serious loss of sites, and the trend continues.

Effects on the perception of nature

The visual impact of afforestation is fundamental to the present issues, because the majority of people who form the nature conservation constituency respond to nature in terms of visual imagery and appeal. Their appreciation of wildlife depends quite closely on the larger setting of habitat and landscape within which it is viewed, so that conservation has to maintain the essential harmony of this relationship. The quality of their experience and enjoyment is thus closely bound up with their perceptions of 'wildness' in nature and is another reason why a tendency to naturalness is highly valued. Wildness is obviously a relative

Significance of afforestation changes to nature conservation

attribute in Britain, where so much of the land has been altered by man, and it tends to be least in the lowlands, in which the greatest development has taken place. Conversely, the uplands are especially valued as Nature's last stronghold — the only places where wild country remains on any scale and the nearest that Britain can show to the spectacular natural areas of other parts of the world. While much of the open ground habitat was deforested long ago and is now only semi-natural, it often lacks the more pervasive signs of recent human activity and so can have an appearance of authentic wild and undisturbed country. It has, over thousands of years, evolved its own distinctive character as wildlife habitat and become an integral part of the British scene, complementing the more intensively managed lands of the enclosed agricultural zone.

Large numbers of people enjoy the new forests for their recreational opportunities, and these places can also have considerable visual qualities where care has been taken over design for scenic amenity and landscape values. Many people nevertheless remain highly critical over modern afforestation in general for its damaging effects on the appearance of the countryside. Within the forests there is typically an enormous contraction in the visual scale of landscape, with wildlife itself being generally far less visible than on open moorland (Moss 1979). So many of the attributes of new forest are in such contrast to those of the ancient semi-natural woodlands that the tree farm analogy is insistently forced on one's awareness, particularly when the forests are vast in area. The wall-like edges to the blocks of trees, the regular rows of trees, the patterns and uniform width of rides, the impenetrability of the thicket stages, the frequent paucity of ground vegetation and the corrugations of ground left by ploughing are unnatural features monotonously present in many forests. Even when there is sympathetic silvicultural treatment, the penetration of each forest by systems of vehicular roads is highly inimical to any sense of naturalness and

The wilderness deer forest country of the western Highlands. Tree-less submontane moorland rising to high and rugged mountains in a region of extreme oceanic climate. Haunt of greenshank and golden eagle.
Strath Lungard, Wester Ross National Scenic Area, Ross-shire, 1966.

Significance of afforestation changes to nature conservation

bird song is too readily interrupted by the sound of a chainsaw. These are, accordingly, additional reasons for regarding the planting even of only a part of an important area as an undesirably intrusive activity.

Future needs for research and information

The ability both to understand the full effects of afforestation on wildlife and environment and to promote improved design and management in future is limited by lack of knowledge and insight. Shortage of resources is a perennial problem, so a strict ordering of priorities is important. Much relevant research in various fields is in hand, but there are still numerous gaps, and new problems tend to appear. The Forestry Research Co-ordinating Committee chaired by FC is a forum for discussion of research needs and the sharing of effort, and NCC is a member.

NCC has for many years been surveying wildlife and its habitats in Britain, to characterise the natural and semi-natural plant communities and measure their extent and to chart the distribution and abundance of plant and animal species. It has commissioned a National Vegetation Classification to provide a reference system for plant communities. The main aims of survey are to identify the areas of greatest importance to nature conservation and the species in greatest need of protection and to set up base-lines for monitoring change to warn of possible conservation problems. Geological and physiographic features have also been surveyed and assessed, with the help of earth scientists outside NCC. Biological areas of National Nature Reserve quality have been described against a classification framework in *A nature conservation review* (Ratcliffe ed. 1977), which has been amended as new key sites have been discovered and others lost. The NGOs, and especially RSPB, have conducted surveys in parallel. In the context of afforestation, NCC has concentrated on surveys of plantable heath, moorland, blanket bog and sand-dunes, to identify SSSI-quality areas. Coverage is still incomplete, especially

Extensive forests on lower moorlands rising into higher hills above the planting limit. The higher, unplantable ground may be important for montane species, but the lower ground often has the highest food value for widely ranging animals such as red deer, raven and golden eagle. Glen Trool Forest and Merrick range, Galloway, 1979.

Significance of afforestation changes to nature conservation

in the Highlands and Islands. The recording of biological attributes for assessment is difficult and slow, but completion of this site assessment and selection programme is regarded as a priority. A Geological Conservation Review is also being undertaken to describe all physical feature sites of national quality.

Knowledge of the losses of habitats, communities and species through afforestation is incomplete and requires a fuller monitoring programme. Knowledge of the biological effects within the new forests is also patchy. NCC has commissioned studies of vegetational change, effects on bird populations and selected species, and the value of some unplanted habitats within forests. However, environmental conditions and the resulting forest successions vary so widely over Britain that a series of studies of forest rotations and residual open habitats needs to be conducted in different locations. Studies of second rotations are only just becoming possible in most areas, as first harvests are being taken. The longer-term dynamics of forest management will require research well into the future. Management options for wildlife need more work, and studies on the habitat needs of individual species are virtually limitless. The priority need is, however, for more experimentation based on present knowledge and the monitoring of such experiments. A discussion of research needs in the new forests is now available (Jenkins ed. 1986), and NCC looks to other interested parties to contribute further, notably the Institute of Terrestrial Ecology, RSPB, the universities, FC and the non-state forestry sector. FC's intentions to expand the scope of its wildlife research are welcomed.

Research into the abiotic environment is technically more demanding and has been done mainly by bodies with particular interest in soils and water such as the Institute of Hydrology, ITE, the Freshwater Fisheries Laboratory of DAFS, other fisheries interests, Water Authorities and universities; FC has also contributed increasingly through its own research. Work in this field probably has a more adequate cover than biological research and is expanding, especially through concern over such matters as acid deposition, water supply and fisheries. The Ordnance Survey maps cannot keep up to date in showing the distribution of afforested ground, so rapidly is this expanding in some districts, but satellite imagery may in future provide more recent information on planting and perhaps also clear-felling. More syntheses of completed research are needed, as are comprehensive reviews of information on ecosystem effects. The forestry interest has a large measure of responsibility for conducting or funding relevant research into the range of afforestation impacts on the environment.

The conflict
of interests
over open
ground

The areas of special importance to nature conservation

When the heaths, grasslands, peatlands or sand dunes that are lost to afforestation have special nature conservation value in their previous state, the forests that replace them, however good they may become for wildlife, are not, and never can be, an adequate substitute. This is so because they can never contain more than a fragmentary, depleted and therefore inadequate representation of the open ground habitats, communities, species and physical features in their former wholeness. Even when only a small part of a special area is affected, new planting introduces an artificial and disruptive element into the ecosystem, disturbing this earlier integrity. There is thus a presumption that proposals for afforestation on SSSIs, using currently favoured species and methods, will be regarded by NCC as damaging operations. This basic incompatibility of interests creates a clash over the special areas when these are subject to proposals for afforestation.

The problem for nature conservation is compounded by the manner of transfer of plantable land to afforestation, which generates a random pressure affecting some of the areas of special importance for wildlife and physical features. Many of the important areas listed on page 55 as damaged or lost were never notified as SSSIs, and some were planted even before the original Nature Conservancy was set up in 1949. In the examples of recent instances of conflict over actual or proposed SSSIs dealt with here, the forestry concerns were aware of the nature conservation interest.

It was not until the mid 1970s, after extensive (though still incomplete) surveys, that NCC notified as SSSIs a bare minimum of large upland and peatland sites, including all the key sites listed in *A nature conservation review*, and undertook further surveys leading to the identification of additional areas of SSSI quality. This has led to an increase, during the last ten years or so, of instances of conflict over forestry proposals on SSSIs. There had

been numerous previous cases of collision of interest on SSSIs, but these were mainly over small sites, of less than 200 ha. Recent cases have been on larger sites, with much more serious implications for both forestry and nature conservation.

The FC is now reviewing its conservation policy but in 1980 it stated: "all designated sites on Commission land are managed to take account of the particular conservation interest concerned." The Commission has also said: "if NCC insists that no planting is permissible, depending on the scale, the FC might accept this 'loss' as part of its conservation commitment or seek to dispose of the area to the NCC" (House of Lords 1980). FC's attitude appears to be influenced by the size of area involved, as evidenced by the concern expressed to NCC over the extent of some upland and peatland SSSIs in Scotland. On the larger SSSIs, the Commission believes that a compromise, allowing some planting, is desirable. In negotiations to establish an NNR at Cairnsmore of Fleet in Galloway, FC was willing to sell or leave unplanted only a part of the area that NCC sought to acquire. FC has set aside a tract of unplanted moorland for its existing wildlife interest in the Kielderhead Conservation Area in the Cheviots, but much of this lies above 450m, the normal planting limit. In Galloway, the leaving of an unplanted wilderness area in the rugged granite hills east of the Merrick has been discussed.

Recent problems have arisen more particularly over non-state forestry, but these also involve FC because of its role in disbursing grant aid. FC consults NCC over planting proposals on SSSIs, or areas which it has been told are proposed SSSIs, but the Commission has to balance the respective weights of all the relevant interests (nature conservation, forestry, recreation, social and economic) and seeks the help of the forestry Regional Advisory Committees in resolving disagreements. If the RACs are unsuccessful, the advice of Ministers is sought. Decisions in favour of planting were made on Mindork Moss, Wigtownshire, and

The conflict of interests over open ground

Arran Northern Mountains SSSIs after call-in by the Secretary of State for Scotland. The best blanket bogs in Wales were recently identified on Llanbrynmair Moors, at the time when a proposal for afforestation arose. The area was in fact planted after a report by an independent arbiter (Lofthouse 1980), though NCC made its importance known to the forestry interest and said that it should be notified as an SSSI.

These three cases occurred in the few years before 1981. Since the passing of the Wildlife and Countryside Act 1981, the only SSSI damaged by afforestation has been Strathy Bogs National Nature Reserve. Planting proposals affecting SSSIs now have to be notified to NCC as a legal requirement and, for the reasons given on p.64, there is a presumption for objections to current afforestation practices as damaging operations. FC has declined grant-aid for three smaller planting applications on SSSIs, but the recent case of Creag Meagaidh has been worrying in its portents.

Part of Creag Meagaidh, a mountain massif in central Inverness-shire, was acquired by a private forestry company in 1983, knowing that it was within an SSSI. FC supported an application for grant-aid to plant part of the area, despite objections by NCC. The RAC could not find a solution acceptable to both parties and the case was referred to the Secretary of State for Scotland, who advised that grant-aid should be given for planting half the proposed area. NCC objected to planting as a damaging operation under the 1981 Act and sought a management agreement under section 28. The outcome was that the land was sold to NCC, at its putative value as plantable hillground. A particular point of concern was that, although Creag Meagaidh had the only significant area of unplanted submontane moorland along a 20-mile stretch of hill country between Spean Bridge and Laggan Bridge, the claim to afforest this remnant was pressed so hard. There appeared, moreover, to be an assumption that, while nature

The advance of afforestation. Intentions for safeguarding the whole of the distant hill in its open condition were unsuccessful and the desired National Nature Reserve boundary was penetrated by blocks of new forest. The open ground in the middle distance has also been planted. Cairnsmore of Fleet NNR, Galloway, 1986.

The conflict of interests over open ground

conservation on this site needed a special justification, forestry did not.

Proposals for planting on another large SSSI in the Perthshire hills were recently rejected by FC in favour of the nature conservation case, which is a most encouraging outcome. Two other cases on large peatland proposed SSSIs in Scotland are currently under consideration.

NCC accepts the principle of compromise as a means of resolving competitive situations, but with an important caveat over how this is interpreted and applied (NCC 1984, section 13.2). The planting of the 1·1 Mha of new forest since 1919 has taken place with local objections and criticisms from scenic amenity interests, but with hardly any opposition from nature conservation interests until quite recent years. NCC itself has raised no objection or even queries to most afforestation, and its recent resistance has been over a very few areas of special importance. Conflict has been avoided in the past because nature conservation has surrendered its interest to afforestation, but there must now be a greater insistence on protecting wildlife values, in the first place by respecting the SSSIs.

NCC believes that recent events have further vindicated the SSSI system, as a countryside network of exemplary areas representing the full range of variation in wildlife habitats and geological and physiographic features. In some districts, these special sites are becoming the only areas which remain protected against transformation by land-use developments such as agriculture and forestry, and any expectation of further compromise within such remaining refuges is unreasonable. The shortcomings of the SSSI system, even since 1981, have been given a considerable airing (e.g. Adams 1986), but one particular weakness is that it is still incomplete. NCC has since 1974 devoted a large part of its very limited resources to survey, but the effort required to obtain environmental, biological and geological information for adequate and consistent evaluation countrywide is so great that the task is still not finished.

Since 1979, particular effort has gone into surveying the lower moorlands and blanket bogs, especially the flow country in the far north-east of Scotland, which is of outstanding international importance (see p.17). Enquiry in the 1970s elicited the view that forestry interest in this flow country was likely to be mainly in the drier moorland and shallower peat areas such as those already planted north of Lairg. Flat bogs had, however, been drained and planted in more southerly districts in the 1960s, and the search for further plantable land inevitably led to the flow country. Forestry interests now own 67,000 ha here, of which 39,000 ha are either already afforested or approved for planting, within a total plantable area of about 190,000 ha. Areas of SSSI quality have been damaged or lost before they could be notified. The total area of existing SSSIs within the flow country is 10,735 ha, but continuing surveys by NCC and RSPB indicate that further areas need to be notified to represent adequately the whole range of nature conservation interest. The pace of developments here and the nature and importance of the issues at stake emphasise that the conservation concern is not simply with the special areas and that preoccupation with the SSSI mechanism has led to a neglect of the seriousness of habitat loss on the wider geographical scale (RSPB 1985; NCC 1986).

The wider countryside

New afforestation has fallen unequally on different parts of Britain (see p.19). The largest individual areas have been planted in Galloway and Carrick (c. 80,000 ha), the Cheviots (c. 60,000 ha), the Kintyre-Knapdale peninsula (c. 55,000 ha), Eskdalemuir and Craik (c. 31,000 ha) and the Breckland (c. 19,500 ha). In these districts most of the plantable land available has been acquired and afforested. This has led to the planting of large continuous blocks creating the appearance of whole landscapes largely covered by trees, and this has been appropriately described as blanket afforestation. When very little of the hill farmland obtained is

The conflict of interests over open ground

Species inhabiting both open ground and new forest
Red deer 1
Black grouse 2
Common aeshna dragonfly 3
Hen harrier 4
Adder 5
Nightjar 6

The conflict of interests over open ground

unplantable, some large, unbroken expanses of plantation can result, for example up to 100km in one block in Kielder Forest. In some districts, planting has formed much more of a patchwork, with blocks of forest as a mosaic with open hill sheepwalk. In high mountain country, forest is restricted to a lower altitudinal zone, with extensive open montane habitat lying above. And in other districts still, there has been relatively little new afforestation at all.

Blanket afforestation over large areas is inimical to nature conservation, even if no specially important areas are planted, since it reduces open ground wildlife and habitats on such a massive scale within a single district. It is not possible to provide detailed, yet comprehensive, balance-sheets of wildlife losses and gains even for a single district, while partial comparisons tend to be selective and biassed, and therefore open to criticism. But in the districts mentioned, so much open ground has now been lost that, whatever the new forests may offer in compensation for the losses, the point of acceptable balance has now been reached or exceeded. Further afforestation in the Breckland appears to have ceased (but see p.34), but in the other four districts, however, it could expand appreciably. Indeed, southern Scotland is regarded by FC as a district suitable for further substantial increase in new planting: 350,000 ha of additional plantable land were identified in Dumfries and Galloway and the Borders Regions in 1980 (House of Lords 1983).

In England and Wales, the National Parks have placed a limited but variable restriction on the spread of new afforestation. The detailed account by MacEwen & MacEwen (1982) shows that National Parks are regarded by forestry interests as areas with a large extent of plantable land (570,000 ha) where considerable further new planting should be allowed. In 1980, over a fifth of FC's estate in England and Wales was in the National Parks, and in the previous year these holdings increased by 8% through acquisitions in the Lake District and Snowdonia alone.

In 1980, FC held 110,666 ha of National Park land, 75% of it planted, and varying from 22% of the Northumberland National Park area to 0·2% of the Yorkshire Dales National Park (mean of 8·1% of total National Parks area). Proposals by the Sandford Committee in 1974 and the Countryside Commission in 1984 that new afforestation in the National Parks and in the uplands of England and Wales respectively should be brought under planning control were rejected by the Government. Attitudes of the Park Authorities to afforestation vary quite widely. Few Areas of Outstanding Natural Beauty have included extensive areas of plantable land, but some of the proposals for new AONBs do so; they also are regarded by foresters as areas within which afforestation should take place.

The Highlands and Islands are probably the area where the forestry interest sees the greatest scope for further expansion (Centre for Agricultural Strategy 1980). National Scenic Areas are evidently not regarded as a constraint (see p.24), and the lower uplands generally are being sought as plantable land, regardless of earlier views on climatic limitations. Even maritime grasslands and heaths which are the feeding habitat of choughs in the west of Islay are under threat of planting, despite exposure to the full force of Atlantic storms. The biggest nature conservation worry is, however, in the flow country of east Sutherland and Caithness, until quite recently seemingly safe as the largest remaining wild and unspoiled peatland area in Britain, though producing a modest return in sheep, deer, grouse, salmon and peat. The need for an appropriate SSSI series here has already been stressed, but there is a distinct prospect that, even if this can be defended, the whole remaining expanse of some 180,000 ha will eventually go under trees. This would be an enormous loss to nature conservation in Britain, in Europe and indeed for the world. The view of the World Conservation Strategy (International Union for Conservation of Nature and Natural Resources 1980)

The conflict of interests over open ground

that the Highlands of Scotland are a priority biogeographical province for the establishment of protected areas is based not only on the intrinsic importance of the region but also on the present insufficiency of such protected areas within it.

Nature conservation and afforestation

7

Future prospects for accommodation between forestry and nature conservation

The economic arguments

The justification given for afforestation taking precedence over nature conservation and other interests is that this is the most productive land-use in the areas concerned and has greater economic validity than the other uses that it replaces. The most familiar form of the argument is that Britain imports 90% of its timber needs, at a cost of £4·5 billion in 1985, so that the need for domestic production to reduce this burden is imperative; NCC must take account of this appeal to the national good because of its general bearing on conservation objectives, NCC's duty to pay regard to the needs of forestry and the economic and social interests of rural areas (see p.73), and the cost of compensation in saving special sites from afforestation.

The economics of forestry are complex and controversial. A government consultative document on forestry policy in 1972 was based on a cost-benefit study by an inter-departmental team of economists under Treasury chairmanship (Treasury 1972). A main conclusion (paragraph 33) was: "the review has shown that the economic return on the Forestry Commission's planting programme leaves a wide gap as compared with either the Government lending rate or the 10 per cent test discount rate which has to be satisfied by a range of other public sector programmes." Earlier (paragraph 10) the review stated: "Looking to the future the Commission expects a net rate of return in real terms for new planting or re-stocking which would range from about 1 per cent on poor sites distant from markets to about 3 per cent on good sites within a reasonable distance of markets. The latter is as high a rate of return as forestry is likely to earn anywhere in the temperate parts of the northern hemisphere." It was proposed that in future the Commission should be set a target rate of return of 3% in real terms, and that a further expansion of new planting on the then current scale could only be justified by the additional social contribution of forestry to rural employment and to amenity and recreational use. Similar conclusions

were drawn for private forestry, but it was noted that the grant and tax arrangements had the probable effect of roughly doubling the 3% rate of return. Private forests were also regarded as much less accessible for public recreation.

This analysis was widely rejected by the forestry interest, which took heart from the industrial growth of the 1970s. The ten per cent discount rate for public sector investment was later considered unrealistically high and reduced to five per cent. *The wood production outlook in Britain* (Forestry Commission 1977) foresaw that, as the 21st century advanced, there would be an increasing likelihood of world timber demands outstripping supply and of price increases relative to other commodities. Further expansion of the national forest estate was considered a prudent policy, though the case for a strategic defence reserve was no longer upheld. A still stronger economic argument was soon presented for a massive increase in new planting, based on the case for import savings, set beside projections of a world timber shortage and price escalations by the end of the century (Centre for Agricultural Strategy 1980).

This economic case has subsequently been challenged, on the grounds that the import savings argument for forestry is doubtful when account is taken of the economic principle of comparative advantage and the need to maintain overseas trade, of the lower costs of timber production in the major exporting countries, and of the opportunity costs of forestry — that is that funds invested in forestry could have been used for alternative enterprises and jobs (Bowers 1984; Grove 1983; Stewart 1985). The forecasts of future wood supply and demand and price increases have already been shown to be substantially overestimated, thereby casting further doubt on their longer-term credibility (Bowers 1984; anon 1985). Stewart (1985) has also suggested that the tropics and subtropics, where climate is optimal for tree growth, could increasingly replace the temperate and Boreal regions as the main centres of softwood timber

Future prospects for accommodation between forestry and nature conservation

production. The job creation aspect of afforestation is also complex. Direct employment in forestry declined from 25,000 + in 1958 to 13,000 + in 1976 through increased mechanisation and productivity, but also through greater use of contract labour, especially for harvesting (CAS 1980). The Forestry Commission employed 6,836 staff in 1984 compared with 14,581 in 1961; and estimates of numbers employed on forestry operations in private forests were 8,000 in 1983 compared with 9,400 in 1961, despite a recent doubling of new planting by the private sector (Hansard 1985a). Extra job creation through expansion of the total forest area thus has to work against the continuing trend of decreasing manpower needs per unit area. The 1972 Treasury study found that forestry gave more employment than farming per unit area of land, but at substantially higher cost per job created. The 'downstream' effect of afforestation on the growth of the wood processing industry is important and likely to increase, but it has to be compared with analogous effects from alternative activities. The degree of integration between forestry and other land-uses also has an important influence on employment aspects. The 1972 Treasury study is widely described as being out of date, but a new and independent examination of the economics of afforestation is urgently needed, giving special attention to opportunity costs, including the 'downstream' aspects, and with more specific recognition of the negative though uncostable effects on nature conservation than in the previous study.

NCC can only note the gist of the economic argument and has to limit comment to aspects especially relevant to nature conservation concern. One main point is that the case for further new afforestation on socio-economic grounds does not carry such an overwhelming and indisputable justification that it should override the needs of nature conservation on sites of special interest for wildlife and physical features. Whether or not forestry is considered to produce a reasonable rate of return for a long-term investment, there are areas from which

we believe the return is higher from an investment in nature conservation. The return from forestry is quantifiable in cash; the return from nature conservation is not, but this is not in itself sufficient justification for commercial criteria to override conservation considerations. Nor does the case appear to have been made for forestry to be allowed an unlimited expansion within any one district. It is certainly an important land-use, but government has recognised the need for it to achieve a balance with other interests in the rural environment. The need to promote economic growth in the remoter regions and to stem rural depopulation is public policy: other activities can, however, help to achieve these socio-economic policy objectives too, and there should be more considered integration between forestry and other rural land-uses. In particular, the encouragement of the combination of farming and forestry on hill farms through well-chosen incentives is advocated as a desirable course and would help to provide employment evenly throughout the year (Stewart 1978; Cunningham *and others* 1978; CAS 1980).

One crucial factor in the economics of afforestation is not in doubt, namely that the whole operation rests on a high level of public subsidy. Various commentators have surmised that very few more forests would be planted in Britain if afforestation had to be run as a normal business operation. The argument for the special nature of forestry, as yielding a long-term return, has always been accepted as the basis for this dependence on public investment. The situation nevertheless requires that full public accountability be satisfied, through the Forestry Commission as the state Forest Authority. At present, "the expected rate of return from private planting is entirely up to the owner who is proposing the private planting" (Hansard 1986). Whatever the economic justification may be for new afforestation, its strength must vary according to an underlying ecological factor, in the widely varying capacity of particular environments to produce quality and quantity of timber. Wood

Nature conservation and afforestation

Future prospects for accommodation between forestry and nature conservation

Species which colonise the new forests
1 Sparrowhawk
2 Foxgloves
3 Goldcrest
4 Pine hawkmoth
5 Chaffinch
6 Male fern

Future prospects for accommodation between forestry and nature conservation

production in Britain clearly ranges from high to low according to differences in conditions for growth (see p.20), and it is measured as timber yield-classes. Much planting is done nowadays on poor land and under adverse climate, where establishment costs are high but timber production is low. While advancing technology is a factor here, the main impetus to afforestation under marginal conditions is the risk protection afforded by state subsidies. Yet it is often in these marginal environments that the damage to nature conservation interest can be high. It is thus especially unacceptable for areas of high interest for their wildlife or physical features to be destroyed in order to produce a timber crop at the bottom end of the yield range.

Given the overall environmental impact of afforestation, critical levels for acceptable ratios between timber production and establishment costs should be set, and below these no state financial support for planting should be allowed. Conversely, there is good ecological sense in encouraging afforestation in lowland, edaphically and climatically favourable environments where timber production will be greatest. Nature conservation value is here often minimal, and conifer plantations could be an advantage. If afforestation could be encouraged on more fertile farmland it might help to overcome present agricultural overproduction and yield more and better timber at lower cost. Such possibilities have been described as 'uneconomic' by private forestry interests (Rankin 1985), because of the high cost of this richer land, but this is an artificial factor which should not be allowed to defeat an otherwise desirable course of action. The declining values of agricultural land may help to ease this problem.

The impact of socio-economic and land-use issues varies from the national down to the individual. NCC operates a countrywide strategy for nature conservation, and the SSSI system has a national perspective, but particular sites involve the interests of particular groups or individuals. While forgoing a specific proposal for planting may be insignificant in terms of national timber production and the economy, it may be highly significant to the people concerned. In areas of high unemployment, the loss of even a few job opportunities created by afforestation can be serious. Indeed, the total effect of nature conservation on the local or regional scale is viewed in some areas, especially of northern Scotland, as an unacceptable limitation on desirable development. NCC is keenly aware of these issues and concerned to approach them positively. A commissioned report (Dartington Institute 1985) has shown that there are at present in Britain around 15,000 full-time jobs (or full-time equivalents) in nature conservation, allowing for indirect categories. NCC is extending this study into further job creation opportunities, especially in remoter rural areas.

The interpretation of reasonable balance

National policies support both the expansion of forestry and the protection of nature, but these policies can be in opposition. Means of resolving the conflicts which arise must be found, but it seems preferable to seek to avoid such conflicts. Both forestry policy and the Forestry Commission are required to achieve an acceptable or reasonable balance between forestry and other environmental interests, and NCC has a duty under the Countryside Act 1968 to have regard to the needs of agriculture and forestry and to the economic and social interests of rural areas. FC is understood to be considering the interpretation of this concept in relation to new afforestation, but its views are not yet known. While this offers a possible way forward, the first step seems to be for the two sides to indicate their objectives as clearly as they are able.

During consultations on *A nature conservation review*, FC indicated to NCC that it regarded about half the area of upland and peatland key sites as plantable land, covering some 200,000 ha. The total area of SSSI land is more than double that of the published NCR site list (see p.11), so that a pro rata increase in plantable land gives about 500,000 ha. This compares with a total extent of plantable land covering about 3 Mha (see p.24). Probably at

Nature conservation and afforestation

7

Future prospects for accommodation between forestry and nature conservation

least 200,000 ha of the plantable land of high quality for nature conservation lies in existing Nature Reserves, National Parks, Areas of Outstanding Natural Beauty and National Trust properties. The proportion of land over which there may be competition from conservation may therefore amount to around 300,000 ha or one tenth of the total plantable area. NCC believes that over this area afforestation should be forgone; existing land-use will mostly continue in such areas and arrangements are available under the 1981 Act to deal with potential financial disadvantage to the landowners concerned. SSSI land is mostly of no special value for afforestation, and much is of low or indifferent productivity.

The situation in the wider countryside particularly focusses on the interpretation of the 'reasonable balance' which the Government feels should be achieved between forestry and other environmental interests. If both forestry and environmental interests were accorded equal weight, the point of balance would be 50% of plantable land to each, but this would be a naive solution. All relevant land-use interests must be involved in the determination of 'reasonable balance'. Farming (especially for sheep and cattle), water users, freshwater fisheries, scenic amenity and other recreational concerns, defence, mineral extractors and local authorities each have a large interest in land suitable for afforestation; and several are represented by both governmental and non-governmental organisations. NCC has itself sought the views of numerous relevant bodies in preparing this document. The nature conservation sector shows a varying degree of coincidence with these other interests, but has a particularly close and supportive affinity with those of scenic amenity and countryside enjoyment, represented officially by the two Countryside Commissions. Another complicating factor over the question of balance is that the previously understood limits of plantable land are presently in a state of flux, particularly because considerable areas of farmed

land seem likely to be surplus in terms of production requirements.

One conclusion is that generalisations about 'reasonable balance' are undesirable and probably unacceptable. It will be a matter of working out an agreed balance — area by area — and the appropriate geographical units are probably counties in England and Wales and regions in Scotland, because of the necessity for the involvement of local authorities. The National Parks and Areas of Outstanding Natural Beauty in England and Wales are special cases, where a good deal of thought has already been given — though seldom with a widely accepted outcome — to the desirable extent of afforestation. The nature conservation view will tend to vary between one county or region and another, according to its perception of the intrinsic merits of the areas concerned. It will be important to identify within the wider countryside the areas where afforestation would have the least damaging impact. All future new afforestation should also meet acceptable environmental standards in general terms.

Nature conservation and scenic/ amenity views on these matters can be expected broadly to coincide, with both regarding further extensive expansion of blanket afforestation as undesirable and recognising a need to restrict new planting in districts with a combination of high wildlife, geological and landscape interest, such as National Parks. The wildlife case will, however, always depend on a detailed evaluation, based on survey data, to map at least broad variations in nature conservation interest over the land concerned. Such evaluations will also be influenced by what has already been lost. Some habitats, such as heather moorland and undisturbed blanket bog, are such a diminishing asset, especially in certain districts, that they will have general priority for retention.

Where natural, or even semi-natural, forest still exists on a large scale it is welcomed by conservationists, and the much lesser extent of open ground habitats and wildlife is then simply accepted as the normal condition. It can thus appear inconsistent for conservationists to place such high value on the large extent of treeless habitats so typical of upland Britain. Some commentators have seen the objections to the visual impact of afforestation as expressing dislike of change and have claimed that, given time and improvement in design and management, the new forests will become accepted by many of their present critics. If the new forests were more natural, they would indeed be more readily accepted, but there is an all-pervading evidence of human presence and intervention which destroys even an illusion of similarity with the real Boreal forests that are part of the northern wilderness regions. But even if the new forests were close to natural, the value of the open moorland and the loss of open ground habitats in the lowlands to intensive agriculture would still lead us to value these habitats sufficiently highly to want to retain reasonable areas of them. The argument is not about keeping forestry out altogether, but about how much there should be in any one district, and how it can avoid the best areas.

On the forestry side, goals appear to be unspecific. Present forestry policy is essentially open-ended, and many foresters would clearly wish it to remain that way. Some, indeed, see a case for maximising domestic wood production by planting as much suitable ground as possible, but the majority would probably accept the Government's viewpoint on the need for compromise (see also p.24). In general, the forestry interest sees new planting continuing as in the past, by the acquisition of hill land as it comes on the market or by persuading landowners of the greater financial benefits of growing trees compared with rearing sheep or game. Afforestation is thus likely to continue expanding according to responses to a market created largely by the extremely favourable tax and grant-aid concessions to non-state forestry. The fact that plantable hill land sold with clearance for forestry commonly fetches three times the price — or even more — that it would command if sold for continuation under sheep-farming

can have only one result — a strong pressure towards afforestation of all such land which comes on the market (Moore 1985b). In this situation, the main factor limiting the rate and scale of new afforestation is the availability of plantable land. Another relevant point is that the large corporate investors forming the secondary market in the private sector are interested mainly in large units of forest, and this promotes blanket rather than integrated afforestation.

NCC accordingly awaits with concern the forestry interest's views on how a reasonable balance between nature conservation and afforestation might be achieved. Pursuit of an open-ended, geographically undefined strategy of expansion is incompatible with any realistic interpretation of balance. And if the case is pressed for planting another $1 \cdot 8$-$2 \cdot 0$ Mha, there is a choice between the past piecemeal approach with an inevitable increase in conflict as the remaining area of plantable land grows smaller or an attempt at accommodation with other interests over an agreed programme. Present arrangements inevitably create a pressure for the planting of any remaining suitable hillground or peatland, so that targets become rather notional.

Finally, NCC is not opposed to further afforestation in the uplands, within an acceptable interpretation of reasonable balance. Its support for it will, however, increase in fairly direct relation to the use of native tree species and the adoption of environmentally sympathetic techniques. Ecologically appropriate reforestation in the right places will often be welcomed. Indeed, while rejecting commercial afforestation on important open ground sites, nature conservationists would sometimes wish to see at least a partial restoration of native woodland cover. Especially where vestiges of such tree cover remain, there is scope for encouraging spread by natural regeneration or by actual planting, or by both. On several National Nature Reserves, NCC has established limited areas of woodland of the once prevailing type and promoted the expansion of existing

75

Future prospects for accommodation between forestry and nature conservation

forest. When commercial return is of lesser importance, it is possible to avoid disruptive methods of ground treatment, to vary species composition (e.g. by using alder and willow on moist seepage areas) and to maintain a patchwork cover, with a good deal of open ground, preserving good examples of existing vegetation. The aim is to simulate the more natural relationships between forest and treeless habitats, so that the wildlife elements of both types coexist as they do in some other countries. The more important open ground habitats will, nevertheless, be left in their treeless state, and on some sites tree regeneration will be actively resisted (e.g. on Breckland grass-heaths, acidic lowland heaths and sand-dune systems).

Possible mechanisms to meet the nature conservation need

Under the present system, the possibilities for achieving special site protection under provisions of the Wildlife and Countryside Act 1981 will depend largely on willingness of the forestry interests to forgo planting opportunities, backed by willingness of government to finance alternative arrangements. Although NCC has voluntarily agreed to extend the compensation principle to forestry (which was not included in the 1981 Act), such compensation cannot cover lost tax allowances, so that the usual option is to buy the land at its value with clearance for planting. However, it is by no means certain that all the special sites can be protected thus, and, in the case of Creag Meagaidh, if the owners had refused to sell and insisted on taking up the grant-aid approved for planting, this would have been acceptable to Ministers and it might not have proved possible to protect the site using a Nature Conservation Order or power of compulsory purchase.

Government policy has already markedly shifted the balance of new planting to the private sector and seeks to minimise controls, emphasising the principles of stewardship amongst landowners and voluntary solutions to problems over competing land-uses. As

Native broadleaved woodland of oak, showing transition to open moorland above: a desirable combination of semi-natural habitats. Haunt of buzzard, wood warbler, red deer and silver-washed fritillary. Horner Valley, Exmoor, Somerset, 1968.

Future prospects for accommodation between forestry and nature conservation

the Government's agency, NCC has striven to make the voluntary principle work and has eschewed calls for planning control over new afforestation, which many others see as the logical, fair and workable solution to these problems. Yet it seems inescapable that some greater degree of regulation of new planting is necessary, if the interests of nature conservation in particular and integrated land-use in general are to be adequately served. The position of private sector afforestation in regard to the level of public subsidy requires some strengthened form of public accountability. The informal controls presently exercised by the Forestry Commission in approving planting proposals should be made mandatory. The idea of planting licences, as recently proposed by the Countryside Commission for Scotland, seems appropriate, but much would depend on the conditions which might be attached to such licences. There should be a presumption that, unless a special case of national interest can be convincingly adduced, damaging proposals to plant SSSI land will be rejected.

Larger afforestation schemes, above a size to be determined by agreement, outside SSSIs should also be subject to formal consultation with the range of land-use interests mentioned on page 74. When there is a clear prospect for massive long-term expansion within a particular district, some form of Environmental Impact Assessment may well be an appropriate mechanism. Local authorities would seem to have a very legitimate interest in new planting, not least because of the emphasis given to its effects on employment, industrial development and related services. The working-out of future land-use programmes for counties and districts, within which the role of forestry would be agreed, seems to NCC a still more desirable course. The Structure Plan approach is well short of strict planning control, and the similar informal approach adjudicated by Mr. R. G. A. Lofthouse over land-use in the Berwyn Mountains (Lofthouse 1980) found favour amongst

the parties concerned. O'Riordan (1983) suggested a three-category classification of rural land, — the Heritage Sites (including SSSIs), which would have a strong presumption against inimical development; Conservation Zones with less outstanding but still important landscape and nature conservation qualities, where development would be regulated by planning control; and Agricultural and Forestry Landscapes in the wider countryside, where there would be a presumption in favour of developments such as afforestation. Whatever route is taken, there is imperative need for early discussion between the various land-use interests and agreement on longer-term consultation procedures. Agriculture Ministers are now considering the designation of a series of Environmentally Sensitive Areas (ESAs); the designation will apply to farmed land, but in view of the increasing pressure to afforest such land, especially in the hills, there may be opportunities within ESAs for the relevant interests to develop a collaborative approach to afforestation programmes.

There seems much merit, in principle, in the afforestation of some lowland agricultural land, as an alternative to excess food production. Not only are there greater opportunities for planting native hardwoods, but even conifer plantations could be managed in ways to give much greater wildlife interest than in many upland situations. Full-length rotations, with heavy thinning regimes, could be a normal choice here, and the scope for diversification in various ways could be considerable. Conifer plantations would, in any case, be likely to develop greater wildlife interest than arable areas. Employment possibilities would increase. Caution is nevertheless necessary over the development of such ideas. They are being actively considered by agricultural and forestry interests, but many practical problems have to be resolved. Other interests besides nature conservation will also wish to be involved. And a great deal rests on precisely which ground is planted. Some believe that the preference will be to plant not the

Future prospects for accommodation between forestry and nature conservation

arable prairies, but the 'in-bye' and marginal lands of the hill farms, or poorer pastures in the lowlands; and this could have a highly undesirable impact on nature conservation. The recent (March 1986) relaxation of agricultural clearance criteria in Scotland will lead to an extension of new planting on to better hill and marginal land, so that forestry will move 'down the hill' though it will not expand on to better grade lower ground, especially arable. The further presumption in favour of forestry in hill areas where the land is unimproved or not capable of making a significant agricultural contribution will facilitate the expansion of afforestation within semi-natural moorland and blanket bog. These changes have quite serious implications for the still wider attrition of upland habitats, yet do nothing to alleviate the problem of farming surpluses in the lowlands of England. This is a *prima facie* indication that the possible future afforestation of good lowland farmland will be regarded as an **addition** and not as a substitute to

present forestry programmes.

All these considerations point to the need for a more definitive statement of government policy on new afforestation. This should recognise that afforestation is a profound change in land-use which is nearly always in competition with other uses and thus cannot be left to an endlessly problematical free-for-all. The present laissez-faire is an archaic and unsatisfactory approach which heightens conflict instead of reducing it. Government should review the needs and claims of the different interests concerned and give more specific guidance than hitherto on how an appropriate balance can be achieved between them. This review should include a careful analysis of all relevant economic and social factors, and it should avoid treating forestry in isolation from other interests. The increasing needs for leisure-time outlets and the potential of tourism must be fully considered. Above all, the future deployment of public subsidies in support of rural land-use should be closely matched to the intended desir-

Fertile pastures and meadows, with higher moorland beyond. A district with little afforestation and outstanding botanical importance in its treeless condition, because of extensive occurrences of unstable, calcareous habitats. Maintenance of traditional upland farming is necessary to retain this interest.
Upper Teesdale NNR, Durham, 1976.

Future prospects for accommodation between forestry and nature conservation

able balance between different activities and should avoid the creation of biases which produce a quite different balance of advantage between interests. The subsidies should also achieve their intended purpose of socio-economic advantage to the rural community.

This analysis of afforestation issues in relation to nature conservation is presented as a contribution to the continuing debate on land-use in Britain. NCC has views on how to deal with some of these problems, but on others it feels that much more discussion with other interests is first necessary. The recommendations which follow are therefore a combination of attempts to work within the existing system and indications of somewhat different solutions to important problems. Above all, NCC appeals to Government to recognise that many of these problems — which will not go away — are integral to the management of the whole rural estate, and so require an integrated rather than a sectional approach. NCC is anxious, in co-operation with others, to play its part in discussions and debate which will lead to desirable solutions.

Nature conservation and afforestation

8

Recommendations: a programme for nature conservation in relation to new afforestation

Action by the Nature Conservancy Council

NCC will pursue the following main objectives in this order of priority.

NCC will need to complete surveys to identify all remaining areas of SSSI quality, followed by notification of these, as quickly as possible. Those sites which would most appropriately be protected under reserve status will be identified. Established procedures for responding to forestry proposals on SSSIs under section 28 of the 1981 Act and for handling management agreements, leases or purchases must be regularly reviewed to ensure that they are working properly. NCC will present its views on conflict cases to Ministers independently of those expressed by the Forestry Commission. The success of the SSSI system will be monitored.

Besides action on SSSIs, NCC will seek ways of avoiding blanket afforestation and will promote the concept of 'reasonable balance' in any one district where afforestation occurs. Such a search will involve discussions with all major land-use interests, and a joint approach with the scenic amenity and countryside organisations. We shall support local approaches towards integrated land-use and shall help to identify areas where afforestation will have the least damaging impact. Within our duty to take account of actual or possible ecological changes we shall maintain a watch over other environmental impacts of afforestation, notably on soils and waters.

NCC will improve its capacity to advise on the design and management of new forests to further nature conservation and will develop appropriate liaison, guidance literature and training courses. It will encourage the forestry interests to establish areas where the principles of such forest management can be studied, demonstrated and monitored.

These objectives will be supported by the following means.

Continuation and expansion of a survey, monitoring, research and information programme. This work will aim at completing the surveys referred to above, and will also extend studies of various aspects (biotic and abiotic) of successional change in the forest rotation and assessment of the overall balance sheet of changes. This programme will take due account of differences along the countrywide gradients of variation in climate, geology and other ecological conditions. Relevant socio-economic studies will also be promoted. A more adequate programme of public information will be developed, involving a range of publications from technical to popular and media presentations.

A programme of liaison with other parties will be developed. This will aim to relate NCC's programme to those of the other organisations concerned, both official and non-official — the forestry and timber organisations, the conservation bodies, agricultural and other land-use interests, and local authorities. On research issues the help of those bodies that have relevant interests will be sought. Funding of research by the developer will also be encouraged.

Action by the Forestry Commission

There are several measures helpful to nature conservation which could be taken within the discretion of present forestry policy and in keeping with the new duty in the Wildlife and Country-side (Amendment) Act 1985 (see p25). The Commission has already said that its management practices seek to achieve a reasonable balance between wood production and nature conservation within the objective to protect and enhance the environment (FC 1980) and it will be valuable if the Commission can more precisely state its interpretation of this duty, and also of the requirement, under the 1980 forestry policy statement (Secretary of State for Scotland), to achieve an acceptable balance between new planting and environmental interests. The following ways are suggested for achieving these aims in practice, subdivided under the Commission's two roles as Forestry Authority and Forestry Enterprise. NCC acknowledges and welcomes the fact that some of these are already established practice.

Recommendations: a programme for nature conservation in relation to new afforestation

Forestry Authority

FC should formally accept the need to protect SSSIs from afforestation that would be damaging to the scientific interest and should decline grant-aid to private forestry interests for planting SSSI land unless NCC advises this would not harm the interest for which the SSSI was notified.

FC should advise government on criteria for estimating acceptable return in timber production as a qualifying threshold for grant-aid, and on how it would propose to regulate the grant-aid function in order to achieve a reasonable balance between afforestation and other land-uses in any county or district.

FC should make adoption of appropriate design and management practices a condition of grant-aid, based on submission of a forest conservation plan (see also recommendations under Forestry Enterprise below). FC should wherever possible encourage detailed liaison with local interests who can advise on the features of environmental value which should be retained within the new forest.

FC should review with NCC the opportunities for forestry staff in public and private sectors to receive training in management for nature conservation.

FC should explore with the Agriculture Departments and other land-use agencies and organisations opportunities for planting enclosed agricultural land on a larger scale than at present, with a view to establishing the changes required to encourage appropriate schemes.

FC's work in disseminating information about good management practice in forestry should be continued and strengthened.

Forestry Enterprise

FC, where it owns or acquires SSSI land, should either manage it to maintain the nature interest or dispose of it to a conservation body. SSSI land should not be afforested unless NCC advises that this is not a damaging operation.

FC should apply the criteria on potential productivity recommended above to its own planting programme. The need to avoid further blanket afforestation in any county or district

and to adopt mosaic planting which concentrates on ground of lowest conservation interest should be followed in FC's own programme according to an understanding on 'reasonable balance' reached with other land-use interests.

FC should further develop and consistently adopt silvicultural practices which minimise adverse impacts on the physical and chemical environment. These include use of planting methods which reduce water run-off and soil erosion, avoidance of ploughing on steep slopes, planting well back from stream banks, reducing fertiliser run-off, and minimising pesticide use.

FC should continue to expand its effort to maximise benefits for wildlife within FC forests, and in particular it should manage its forests in accordance with a forest conservation plan agreed with NCC, taking account of the measures listed in chapter 5 (pp.46-47). Experimental conservation management areas should be established and monitored. Forestry research should contribute to these ends.

FC research (either 'in-house' or contracted out) should develop a greater input to wildlife conservation studies. There should be a greater contribution to surveys of the wildlife value and changes within forests, and on unplanted adjoining or included ground, and to management-orientated work on both plants and animals. FC should review the opportunities for addressing the environmental implications of plantation design and management in its Research Branch's programme and consult NCC on issues requiring joint attention. Forest rangers' training should include wildlife and nature conservation aspects.

Action by the private forestry sector

Most of the above points apply equally to the non-state forestry sector, and NCC hopes that they can be accepted, through the good offices of the Timber Growers UK, owners and occupiers, forestry companies and the forestry profession. The following additional action is suggested.

Promotion of guidelines for

Recommendations: a programme for nature conservation in relation to new afforestation

foresters on practical ways to minimise adverse effects of afforestation and to enhance opportunities for wildlife in plantations (see also p.81). A valuable start has been made by TGUK with their *Forestry and woodland code* (1985).

Contributions to a programme of research on the environmental effects of forestry and the management of forests for wildlife.

Discussion with others to assess further opportunities for forestry in relation to agriculture (see also p.81).

Support for the Farming, Forestry and Wildlife Advisory Groups and help to enlarge the forestry-orientated activities of these groups.

Action by government

Achieving a better relationship between nature conservation and afforestation requires a combined approach from all concerned. Yet a situation in which various sectional interests — all purporting to represent a national interest — are competing for a common resource of land has endless possibilities for uncertainty and conflict unless higher authority helps to reach a state of resolution. It is not enough for government to give general admonitions on achieving reasonable balance and having regard to the interests of other land-users: clear principles, guidelines and mechanisms for achieving such appropriate sharing of the common resource are needed. There is thus a great need for central policy makers to grasp the nettle and decide how to achieve the appropriate solutions.

After discussions with all relevant departments and their statutory advisers, government should therefore publish a Green Paper on the rural estate of Britain, with a reconsideration of national forestry policy as one of the central issues. The need is for a more definitive policy embodying an agreed programme with specified targets, in place of the present haphazard and open-ended expansion of afforestation, and for a system of incentives specifically geared to the defined purpose. The objectives and economics of forestry should be re-examined, paying regard to the recommendations of the Public

Accounts Committee on forestry taxation (House of Commons 1980). One essential particular adjustment is to remove the automatic grant of tax relief for all planting schemes, whether or not approved for FC grant-aid. This could be done by instituting a system of planting licences administered by FC. Such licences should be required for schemes above a certain size limit (say 10 ha) and there should be qualifying standards in terms of estimated productivity; there should also be some form of indicative planting strategy to guide the issue of licences, region by region. Tax relief would thus apply only to licensed schemes and to those exempt on grounds of size.

One result of past and present forestry policies and the financial incentives supporting them has been the separation of forestry from agriculture. While some independence between the forestry and agriculture sectors of our economy may be necessary, there is also a need to make these two activities more mutually supportive in terms of land-use, advice, employment and financial returns to owners and occupiers. Present policies benefit wealthy individual or institutional investors and the larger landowners, but they have done little to encourage the smaller farmer to practise forestry. This underlines the need to review the tax incentives for forestry, which provide little benefit to farmers who are not high income earners. Government needs to provide this encouragement, which should thereby help to achieve a more effective integration between these two land-uses, promote the objective of 'reasonable balance' and achieve increased employment in the countryside. The promotion of forestry in lowland agricultural areas will need careful design and close liaison between the two interests, with defined procedures, objectives and criteria for such transfer of use. The matter has been considered in the report on the scope for encouraging wood as a farm crop (MAFF 1985).

There is a need to move away from the narrowly timber-oriented Forestry Grant Scheme, to arrangements which

Recommendations: a programme for nature conservation in relation to new afforestation

would provide for a wider range of objectives, as envisaged under the more flexible provisions of Article 20 of the EEC Regulations on improving the efficiency of agricultural structures.

Representation of interests on Regional Advisory Committees of the Forestry Commission should contain a more favourable balance of environmental and public concerns (in relation to those favouring timber-growing and land-owning interests), and their proceedings should be made more public. FC's recent proposals for changes in the composition and procedures of RACs appear to go some way towards meeting this point and are therefore welcomed.

Action by local government

As afforestation constitutes such a radical change in land-use, local authorities should seek to enlarge their involvement in it, not only in regard to amenity issues but also throughout the full range of their responsibilities. The initiatives of some local authorities in zoning maps with presumptions for and against forestry could be usefully extended. The development of greater 'in-house' technical expertise on forestry and other interests would be beneficial to such increased involvement in rural land-use. Efforts by National Park Authorities to obtain a programmed statement of forestry intentions in Park Plans should be supported.

Action by the NGOs for nature conservation

The NGOs are the private sector for nature conservation, and their independent status equips them to promote the nature conservation case in ways which both support and complement the work of NCC. They should stimulate informal debate on forestry policy, supplying accurate information to back the wildlife case, and provide a channel for the expression of public opinion to policy-makers. They should disseminate information to the rest of society about afforestation issues and the need for greater attention and support for nature conservation in relation to these. They should contribute as far as possible to the site safeguard programme, in close

consultation with NCC. In particular, RSPB is well placed to set up large upland reserves in districts subject to afforestation. RSPB also works closely with NCC on relevant survey, monitoring and research, and the rest of the NGOs contribute to recording and monitoring schemes (eg the mapping schemes of the Botanical Society of the British Isles, British Trust for Ornithology, etc), and this input could become still more valuable and effective through increased organisation. NGOs should also provide advice to foresters on the design and management of new and existing forests. Nature Conservation Trusts and Farming, Forestry and Wildlife Advisory Groups have a special role in advising farmers who wish to convert areas of their farms to forestry.

Action by other environmental interests

Other bodies have their own concerns and approaches over afforestation, and the main need is for these to be developed further and with as much collaboration as is appropriate. Each should be prepared to contribute its views to the interpretation of 'reasonable balance' which FC is now examining. The Countryside Commission has recently expressed its views on afforestation. The Countryside Commission for Scotland is currently considering the subject and will presumably take a policy view, including a view on the role of forestry within National Scenic Areas. The NGOs for scenic amenity and countryside recreation have mostly declared their attitudes. It would be helpful if these bodies could integrate the nature conservation view with their own, so that the two become more mutually supportive than in the past. Their co-operation with the nature conservation organisations in presenting a collective view whenever possible would be welcome. Freshwater and fishing interests could also usefully exchange views and information with the nature conservation organisations.

Conclusion

Forestry is the next most important rural land-use in Britain after agriculture. Indeed, in some upland districts it has become the dominant land-use, and its importance increases steadily. As with agriculture, the 20th century has seen the development of a paradoxical situation whereby a traditional activity which once tended to sustain or even enrich the capital of wild nature has become steadily more inimical in its effects on this resource. The restoration of long-lost woodland cover to open landscapes has not received from nature conservationists the unqualified welcome which might have been their expected response. This review has tried to set out the precise reasons for this state of affairs. Broadly, they are, as in farming, related to the increasing modernisation of techniques in the direction of high input/high output systems. In nature conservation terms, forestry spans a range from the completely benign and desirable to the totally incompatible and unwanted. In the case of the long-established native woodlands, a willingness to understand the problems and work towards reasonable solutions has within a few years shown promise of a much closer and more mutually supportive relationship between forestry and nature conservation interests. The issue of new afforestation has yet to be similarly resolved, but the nature conservation side believes that in this matter, too, where there is a will there is a way.

ADAMS, W. 1986.
Nature's place: conservation sites and countryside change.
London, Allen and Unwin. (In press).

ANDERSEN, S. T. 1969.
Interglacial vegetation and soil development.
Geological Society of Denmark. Bulletin, 19, 90-102.

ANDREWS, J., BAINBRIDGE, I., & BROOKE, C. 1985.
Forestry and bird conservation in the United Kingdom.
Sandy, Royal Society for the Protection of Birds. (Conplan Topic Paper 13).

ANON. 1985.
Coniphobia.
Economist, Sept. 7, 36.

BALFOUR, W. J. 1981.
National Scenic Areas and their implications for forestry in the Northern Region.
Scottish Forestry, 35, 289-294.

BATTARBEE, R. W., and others. 1985.
²¹⁰Pb dating of Scottish lake sediments, afforestation, and accelerated soil erosion.
Earth Surface Processes and Landforms, 10, 137-142.

BIBBY, C. J., PHILLIPS, B. N., & SEDDON, A. J. E. 1985.
Birds of restocked conifer plantations in Wales.
Journal of Applied Ecology, 22, 619-633.

BINNS, W. O. 1979.
The hydrological impact of afforestation in Great Britain.
In: Man's impact on the hydrological cycle in the United Kingdom, ed. by G. E. Hollis, 55-69. Norwich, Geo Abstracts.

BLACKIE, J. R., & NEWSON, M. D.
The effects of forestry on the quantity and quality of runoff in upland Britain. (In press).

BORMANN, F. H., *and others.* 1974.
The export of nutrients and recovery of stable conditions following deforestation at Hubbard Brook.
Ecological Monographs, 44, 255-277.

BOWERS, J. K. 1984.
Do we need more forests?
Leeds, Leeds University School of Economics. (Discussion Paper Series No. 137).

CALDER, I. R., and NEWSON, M.D. 1979.
Land use and upland water resources in Britain — a strategic look.
Water Resources Bulletin, 15, 1628-1639.

CALDER, I. R., & NEWSON, M. D. 1980.
The effects of afforestation on water resources in Scotland.
In: Land assessment in Scotland: proceedings, Royal Scottish Geographical Society Symposium, Edinburgh, 1979, ed. by M. F. Thomas, & J. T. Coppock, 51-62. Aberdeen, Aberdeen University Press.

CENTRE FOR AGRICULTURAL STRATEGY. 1980.
Strategy for the UK forest industry.
Reading. (CAS Report No. 6).

CLARKE, R. T., & McCULLOCH, J. S. G. 1979.
The effect of land use on the hydrology of small upland catchments.
In: Man's impact on the hydrological cycle in the United Kingdom, ed. by G. E. Hollis, 71-78. Norwich, Geo Abstracts.

CLARKE, W. G. 1937.
In Breckland wilds. 2nd ed., rev. by R. R. Clarke. Cambridge, Heffer.

COULSON, C. B., DAVIES, R. I., & LEWIS, D. A. 1960.
Polyphenols in plant, humus, and soil. I. Polyphenols of leaves, litter, and superficial humus from mull and mor sites.
Journal of Soil Science, 11, 20-29.

COUNTRYSIDE COMMISSION. 1984.
A better future for the uplands.
Cheltenham. (CCP 162).

COY, J. S. 1979.
Forestry Commission aerial application of fertilisers: effects on water quality.
Dumfries, Solway River Purification Board. (Unpublished report).

CRAWFORD, D. B. 1979.
Forestry in the national economy.
Scottish Forestry, 33, 29-36.

CRICK, H. Q. P. 1986.
Effects on non-target animals of insecticide applications to control pine beauty moths in commercial plantations in Scotland.
In: Symposium on trees and wildlife in the Scottish uplands, ed. by D. Jenkins. Abbots Ripton, Institute of Terrestrial Ecology. (Symposium No. 17). (In press).

CUNNINGHAM, J. M. M., *and others.* 1978.
Inter-relations between agriculture and forestry: an agricultural view.
Scottish Forestry, 32, 182-193.

CURRIE, F. A., & BAMFORD, R. 1981.
Bird populations of sample pre-thicket forest plantations.
Quarterly Journal of Forestry, 75, 75-82.

CURRIE, F. A., & BAMFORD, R. 1982.
Songbird nestbox studies in forests in North Wales.
Quarterly Journal of Forestry, 76, 250-255.

DARTINGTON INSTITUTE. 1985.
Employment and nature conservation.
Peterborough, Nature Conservancy Council.
(CSD Report No. 617).

DAVIES, E. J. M. 1978.
The future development of even-aged
plantations: management implications.
In: The ecology of even-aged forest plantations
ed. by E. D. Ford, D. C. Malcolm & J. Atterson,
465-480. Cambridge, Institute of Terrestrial
Ecology.

DAVIES, R. I., COULSON, C. B., & LEWIS, D. A.
1964.
Polyphenols in plant, humus, and soil. IV. Factors
leading to increase in biosynthesis of
polyphenols in leaves and their relationship to
mull and mor formation.
Journal of Soil Science, 15, 310-318.

DEPARTMENT OF THE ENVIRONMENT *and
others.* 1982.
*Sites of Special Scientific Interest: code of
guidance.*
London, HMSO.

DRAKEFORD, T. 1979.
*Report of survey of the afforested spawning
grounds of the Fleet catchment.*
Dumfries, Forestry Commission South Scotland
Conservancy. (Unpublished report).

DRAKEFORD, T. 1982.
*Management of upland streams (an experimental
fisheries management project on the afforested
headwaters of the River Fleet,
Kirkcudbrightshire): paper presented to Institute
of Fisheries Management, 12th Annual Study
Course, Durham.*

EGGLISHAW, H. J. 1985.
Afforestation and fisheries.
*In: Habitat modification and freshwater fisheries:
proceedings of a symposium of the European
Inland Fisheries Advisory Commission,* ed. by J.
S. Alabaster, 236-244. London, Butterworth.

FLOWER, R. J., & BATTARBEE, R. W. 1983.
Diatom evidence for recent acidification of two
Scottish lochs.
Nature, 305, 130-133.

FORD, E. D., MALCOLM, D. C., & ATTERSON, J.
eds. 1978.
*The ecology of even-aged forest plantations:
proceedings of the meeting of Division I,
International Union of Forestry Research
Organisations, Edinburgh.*
Cambridge, Institute of Terrestrial Ecology.

FORESTRY COMMISSION. 1977.
The wood production outlook in Britain: a review.
Edinburgh.

FORESTRY COMMISSION. 1980 (rev. 1986).
The Forestry Commission and conservation.
Edinburgh. (Policy and Procedure Paper No. 4).

FORESTRY COMMISSION. 1983a.
*The use of chemicals (other than herbicides) in
forestry and nursery.*
Edinburgh. (Booklet No. 52).

FORESTRY COMMISSION. 1983b.
The use of herbicides in the forest.
Edinburgh. (Booklet No. 51).

FORESTRY COMMISSION. 1984.
*Census of woodlands and trees 1979-82: Great
Britain.*
Edinburgh.

FORESTRY COMMISSION. 1985.
Forestry facts and figures.
Edinburgh.

FRY, G. L. A., & COOKE, A. S. 1984.
*Acid deposition and its implications for nature
conservation in Britain.*
Shrewsbury, Nature Conservancy Council.
(Focus on Nature Conservation No. 7).

GASH, J. H. C., & STEWART, J. B. 1977.
The evaporation from Thetford Forest during
1975.
Journal of Hydrology, 35, 385-396.

GIBSON, C. E. 1976.
An investigation into the effects of forestry
plantations on the water quality of upland
reservoirs in Northern Ireland.
Water Research, 10, 995-998.

GODDARD, T. R. 1935.
A census of short-eared owls (*Asio f. flammeus*) at
Newcastleton, Roxburghshire, 1934.
Journal of Animal Ecology, 4, 113-118, 289-290.

GRAESSER, N. W. 1979.
Effect on salmon fisheries of afforestation, land
drainage and road making in river catchment
areas.
Salmon Net, 12, 38-45.

GRIBBLE, F. C. 1983.
Nightjars in Britain and Ireland in 1981.
Bird Study, 30, 165-176.

GROVE, R. 1983.
*The future for forestry: the urgent need for a new
policy.*
Cambridge, British Association of Nature
Conservationists.

HAMILTON, G. A., & RUTHVEN, A. D. 1981.
The effects of fenitrothion on some terrestrial
wildlife.
*In: Aerial application of insecticide against pine
beauty moth,* ed. by A. V. Holden, & D. Bevan,
91-100. Edinburgh, Forestry Commission.
(Occasional Paper No. 11).

HANSARD. 1985a.
Forestry employment.
Hansard (HC), Written Answers, May 8, 445.

HANSARD. 1985b.
Forestry Commission (Aerial spraying).
Hansard (HC), Written Answers, May 22, 430.

HANSARD. 1986.
Afforestation.
Hansard (HC), Oral Answers, February 26, 936.

HARRIMAN, R. 1978.
Nutrient leaching from fertilized forest
watersheds in Scotland.
Journal of Applied Ecology,
15, 933-942.

HARRIMAN, R., & MORRISON, B. R. S. 1982.
Ecology of streams draining forested and non-
forested catchments in an area of central
Scotland subject to acid precipitation.
Hydrobiologia, 88, 251-263.

HARRIMAN, R., & WELLS, D. E. 1985.
Causes and effects of surface water acidification
in Scotland.
Water Pollution Control, 84, 215-224.

HARRIS, J. A. 1983.
Birds and coniferous plantations.
Tring, Royal Forestry Society.

HELLIWELL, D. R. 1978.
Forestry's long term environmental role.
In: The future of upland Britain, ed. by R. B.
Tranter, 108-113. Reading, Centre for Agricultural
Strategy. (Strategy Paper No. 2).

HIBBERD, B. G. 1985.
Restructuring of plantations in Kielder Forest
District.
Forestry, 58, 119-129.

HILL, M. O. 1979.
The development of a flora in even-aged
plantations.
In: The ecology of even-aged forest plantations,
ed. by E. D. Ford, D. C. Malcolm, & J. Atterson,
175-192.
Cambridge, Institute of Terrestrial Ecology.

HILL, M. O. 1983.
Plants in woodlands.
*In: Centenary conference on forestry and
conservation,* ed. by E. H. M. Harris, 56-68.
Tring, Royal Forestry Society.

HOLMES, G. D. 1979.
An introduction to forestry in upland Britain.
In: The ecology of even-aged forest plantations,
ed. by E. D. Ford, D. C. Malcolm, & J. Atterson,
7-19.
Cambridge, Institute of Terrestrial Ecology.

HOUSE OF COMMONS. COMMITTEE OF
PUBLIC ACCOUNTS. 1980.
Eighth report Inland Revenue: taxation.
London, HMSO. (HCP 448, Sess. 1979-80).

HOUSE OF LORDS. SELECT COMMITTEE ON
SCIENCE AND TECHNOLOGY. 1980.
*Scientific aspects of forestry: Vol. I: report; Vol. II:
minutes of evidence.*
London, HMSO. (HLP 381, Sess. 1979-80).

HOUSE OF LORDS. SELECT COMMITTEE ON
SCIENCE AND TECHNOLOGY. 1983.
*Guidelines on land use: Government response to
the fourth report of the Select Committee (Session
1981-82) on the scientific aspects of forestry.*
London, HMSO. (HLP 11, Sess. 1983-84).

INTERNATIONAL UNION FOR
CONSERVATION OF NATURE AND NATURAL
RESOURCES. 1980.
*World conservation strategy; living resource
conservation for sustainable development.*
Gland, Switzerland.

JACK, W. L. 1980.
*Forestry and water yield: the water industry
position.*

JENKINS, D. ed. 1986.
*Symposium on trees and wildlife in the Scottish
uplands.*
Abbots Ripton, Institute of Terrestrial Ecology.
(Symposium No. 17). (In press).

LACK, D. 1933.
Habitat selection in birds with special reference
to the effects of afforestation on the Breckland
avifauna.
Journal of Animal Ecology, 2, 239-262.

LACK, D. 1939.
Further changes in the Breckland avifauna
caused by afforestation.
Journal of Animal Ecology, 8, 277-285.

LAW, F. 1956.
The effect of afforestation upon the yield of water
catchment areas.
British Waterworks Association. Journal, Nov.,
489-494.

LESLIE, R. 1981.
Birds of the North East England forests.
Quarterly Journal of Forestry, 75, 153-158.

LESLIE, R., & HOBLYN, R. 1984.
Birds of Thetford Forest.
Cambridge, Forestry Commission East England
Conservancy. (Unpublished report).

LOFTHOUSE, R. G. A. 1980.
*The Berwyn mountains area of Wales: an
appraisal.*
Reading, College of Estate Management.

MacEWEN, A., & M. 1982.
National parks: conservation or cosmetics?
London, Allen and Unwin.

MARQUISS, M., NEWTON, I., &
RATCLIFFE, D. A. 1978.
The decline of the raven *Corvus corax* in relation
to afforestation in southern Scotland and northern
England.
Journal of Applied Ecology, 15, 129-144.

MARQUISS, M., & NEWTON, I. 1982.
The goshawk in Britain.
British Birds, 75, 243-259.

MARQUISS, M., RATCLIFFE, D. A., &
ROXBURGH, R. 1985.
The numbers, breeding success and diet of
golden eagles in southern Scotland in relation to
changes in land use.
Biological Conservation, 34, 121-140.

MEARNS, R. 1983.
The status of the raven in southern Scotland and
Northumbria.
Scottish Birds, 12, 211-218.

MILES, J. 1978.
The influence of trees on soil properties.
*Institute of Terrestrial Ecology. Annual Report,
1977,* 7-11.

MILES, J. 1986.
What are the effects of trees on soils?
*In: Symposium on trees and wildlife in the
Scottish uplands,* ed. by D. Jenkins.
Abbots Ripton, Institute of Terrestrial Ecology.
(Symposium No. 17). (In press).

MILLS, D. H. 1980a.
The management of forest streams.
London, HMSO. (Forestry Commission Leaflet
No. 78).

MILLS, D. H. 1980b.
Scottish salmon rivers and their future
management.
In: Atlantic salmon: its future, ed. by A. E. J. West,
70-81. Farnham, Fishing News Books.

MINISTRY OF AGRICULTURE, FISHERIES AND
FOOD *and others.* 1985.
*Report of a working group set up to examine the
scope for encouraging wood as a farm crop.*
London.

MOORE, P. J. 1985a.
The real world of private forestry.
Ecos, 6(2), 2-7.

MOORE, P. J. 1985b.
The unacceptable face of private forestry.
Ecos, 6(4), 34-40.

MOSS, D. 1978.
Song-bird populations in forestry plantations.
Quarterly Journal of Forestry, 72, 5-14.

MOSS, D. 1979.
Even-aged plantations as a habitat for birds.
In: The ecology of even-aged forest plantations,
ed. by E. D. Ford, D. C. Malcolm, & J. Atterson,
413-427. Cambridge, Institute of Terrestrial
Ecology.

NATURE CONSERVANCY COUNCIL. 1983.
*SSSIs: what you should know about Sites of Special
Scientific Interest.*
London.

NATURE CONSERVANCY COUNCIL. 1984.
Nature conservation in Great Britain.
London.

NATURE CONSERVANCY COUNCIL. 1986.
The afforestation of peatlands in Sutherland and
Caithness.
NCC Topical Issues, 2(1), 4-6.

NETHERSOLE-THOMPSON, D., & M. 1979.
Greenshanks.
Berkhamsted, Poyser.

NEWSON, M. 1985.
Forestry and water in the uplands of Britain: the
background of hydrological research and
options for harmonious land-use.
Quarterly Journal of Forestry, 79, 113-120.

NEWTON, I. 1972.
Finches.
London, Collins. (New Naturalist Series No. 55).

NEWTON, I. 1984.
Upland forestry brings wildlife gains.
Economic Forestry Group Magazine, 1984, 8-10.

NEWTON, I., MEEK, E. R., & LITTLE, B. 1978.
Breeding ecology of the merlin in
Northumberland.
British Birds, 71, 376-398.

OGILVIE, R. S. D., & LAMB, R. 1986.
Whither forestry? The scene in AD 2025: a
forester's view.
*In: Symposium on trees and wildlife in the
Scottish uplands,* ed. by D. Jenkins. Abbots
Ripton, Institute of Terrestrial Ecology.
(Symposium No. 17). (In press).

O'RIORDAN, T. 1983.
Putting trust in the countryside.
*In: The conservation and development
programme for the UK: a response to the World
Conservation Strategy,* 171-260. London,
Kogan Page.

ORMEROD, S. J., TYLER, S. J., & LEWIS, J. M. S.
1985.
Is the breeding distribution of dippers
influenced by stream acidity?
Bird Study, 32, 32-39.

PEARSALL, W. H. 1938.
The soil complex in relation to plant communities.
III. Moorlands and bogs.
Journal of Ecology, 26, 298-315.

PEARSALL, W. H. 1950.
Mountains and moorlands.
London, Collins. (New Naturalist Series No. 11).

PEGLAR, S. 1979.
A radiocarbon-dated pollen diagram from Loch
of Winless, Caithness, north-east Scotland.
New Phytologist, 82, 245-263.

PRESTT, I. 1985.
Comment: forestry threatens flow country.
Birds, 10(6), 5.

PYATT, D. G., & CRAVEN, M. M. 1979.
Soil changes under even-aged plantations.
In: The ecology of even-aged forest plantations,
ed. by E. D. Ford, D. C. Malcolm, & J. Atterson,
369-386.
Cambridge, Institute of Terrestrial Ecology.

RANKIN, K. 1985.
Jobs from trees.
Daily Telegraph, Dec. 20.

RATCLIFFE, D. A. ed. 1977.
A nature conservation review.
Cambridge, Cambridge University Press.

RATCLIFFE, D. A. 1980.
The peregrine falcon.
Calton, Poyser.

RATCLIFFE, D. A. 1986.
The effects of afforestation on the wildlife of open
habitats.
*In: Symposium on trees and wildlife in the
Scottish uplands,* ed. by D. Jenkins. Abbots
Ripton, Institute of Terrestrial Ecology.
(Symposium No. 17). (In press).

REED, T. 1982.
Birds and afforestation.
Ecos, 3(1), 8-10.

ROBERTS, G. 1985.
*Nutrient losses from upland catchments: paper
given to Institute of British Geographers annual
conference, Leeds, January 1985.*
(Unpublished paper).

ROBINSON, M. 1980.
*The effect of pre-afforestation drainage on the
streamflow and water quality of a small upland
catchment.*
Wallingford, Institute of Hydrology.
(Report No. 73).

ROBINSON, M., & BLYTH, K. 1982.
The effect of forestry drainage operations on
upland sediment yields: a case study.
Earth Surface Processes and Landforms, 7, 85-90.

ROBINSON, M., & NEWSON, M. D. 1986.
Peat hydrology under moorland and forest on
Plynlimon, mid-Wales.
International Peat Society. Bulletin. (In press).

ROYAL SOCIETY FOR THE PROTECTION OF
BIRDS. 1985.
Forestry in the flow country — the threat to birds.
Sandy.

SANDFORD, *Lord.* Chairman. 1974.
*Report of the National Park Policies Review
Committee.*
London, HMSO.

SECRETARY OF STATE FOR SCOTLAND. 1980.
Statement on forestry policy review.
Hansard (HC), Fifth Series, 995, 927-935.

SMITH, B. D. 1980.
The effects of afforestation on the trout of a small
stream in southern Scotland.
Fish Management, 11, 39-58.

STAINES, B. W. 1983.
The conservation and management of mammals
in commercial plantations with special reference
to the uplands.
*In: Centenary conference on forestry and
conservation,* ed. by E. H. M. Harris, 38-55.
Tring, Royal Forestry Society.

STEELE, R. C. 1972.
Wildlife conservation in woodlands.
London, HMSO. (Forestry Commission Booklet
No. 29).

STEELE, R. C., & BALFOUR, J. 1980.
Nature conservation in upland forestry —
objectives and strategy.
In: Forestry and farming in upland Britain, 161-192.
Edinburgh, Forestry Commission. (Occasional
Paper No. 6).

STEWART, G. G. 1978.
Inter-relations between agriculture and forestry
in the uplands of Scotland: a forestry view.
Scottish Forestry, 32, 153-181.

STEWART, L. 1963.
*Investigation into migratory fish propagation in
the area of the Lancashire River Board.*
Lancaster, Barker.

STEWART, P. J. 1985.
British forestry policy: time for a change?
Land Use Policy, 2, 16-29.

STOAKLEY, J. T. 1986.
Protecting the timber resource.
*In: Symposium on trees and wildlife in the
Scottish uplands,* ed. by D. Jenkins.
Abbots Ripton, Institute of Terrestrial Ecology.
(Symposium No. 17). (In press).

STONER, J. H., GEE, A. S., & WADE, K. R. 1984.
The effects of acidification on the ecology of
streams in the upper Tywi catchment in
West Wales.
Environmental Pollution, A, 35, 125-157.

STONER, J. H., & GEE, A. S. 1985.
Effects of forestry on water quality and fish in
Welsh rivers and lakes.
*Institute of Water Engineers and Scientists.
Journal, 39,* 27-45.

STRETTON, C. 1984.
Water supply and forestry — a conflict of
interests: Cray Reservoir, a case study.
*Institute of Water Engineers and Scientists.
Journal, 38,* 323-330.

STROUD, D. A., & REED, T. M. 1986.
The effect of plantation proximity on moorland
breeding waders.
Wader Study Group. Bulletin, 44. (In press).

Nature conservation and afforestation 10

References

SWANK, W. T., & DOUGLASS, J. E. 1974.
Streamflow greatly reduced by converting deciduous hardwood stands to pine.
Science, 185, 857-859.

TANSLEY, A. G. 1939.
The British Islands and their vegetation.
Cambridge,Cambridge University Press.

THOMPSON, D. B. A., THOMPSON, P. S., & NETHERSOLE-THOMPSON, D. 1986.
Timing of breeding and breeding performance in a population of greenshanks *(Tringa nebularia).*
Journal of Animal Ecology, 55, 181-199.

TIMBER GROWERS UNITED KINGDOM. 1985.
The forestry and woodland code.
London.

TREASURY. 1972.
Forestry in Great Britain: an inter-departmental cost/benefit study.
London, HMSO.

VILLAGE, A. 1982.
The home range and density of kestrels in relation to vole abundance.
Journal of Animal Ecology, 51, 413-428.

WALSH, P. D. 1980.
The impacts of catchment afforestation on water supply interests.

WARREN SPRING LABORATORY. 1983.
Acid deposition in the United Kingdom.
Warren Spring Laboratory, Stevenage.

WATSON, A. D. 1977.
The hen harrier.
Berkhamsted, Poyser.

WATT, A. D. 1986.
The ecology of the pine beauty moth in commercial woods in Scotland.
In: Symposium on trees and wildlife in the Scottish uplands, ed. by D. Jenkins.
Abbots Ripton, Institute of Terrestrial Ecology.
(Symposium No. 17). (In press).

WILDLIFE CONSERVATION SPECIAL COMMITTEE (ENGLAND AND WALES). 1947.
Conservation of nature in England and Wales.
London, HMSO. (Cmd. 7122).

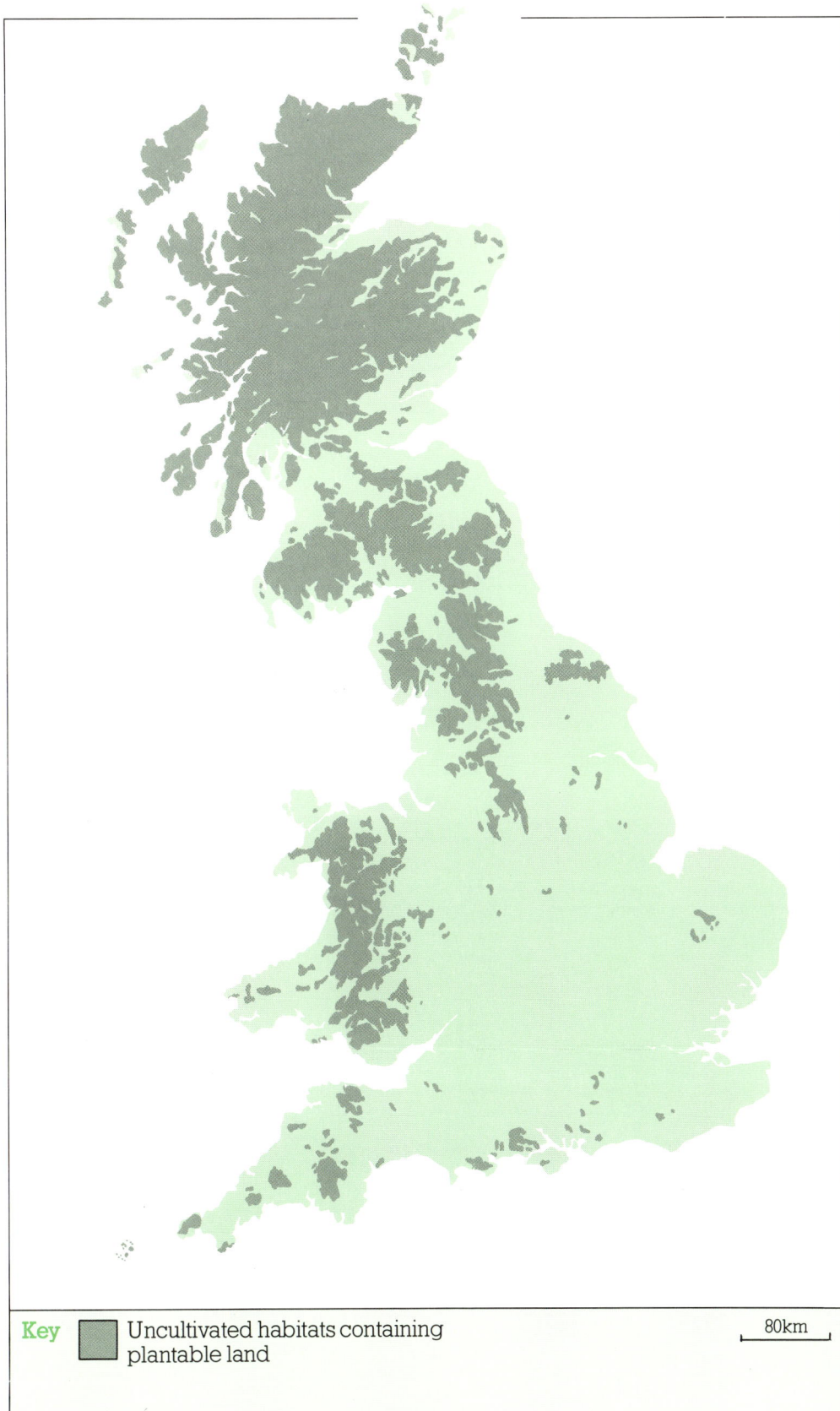

Key — Uncultivated habitats containing plantable land

80km

91

Maps and tables

Map 2 The distribution of existing forest planted since 1920.

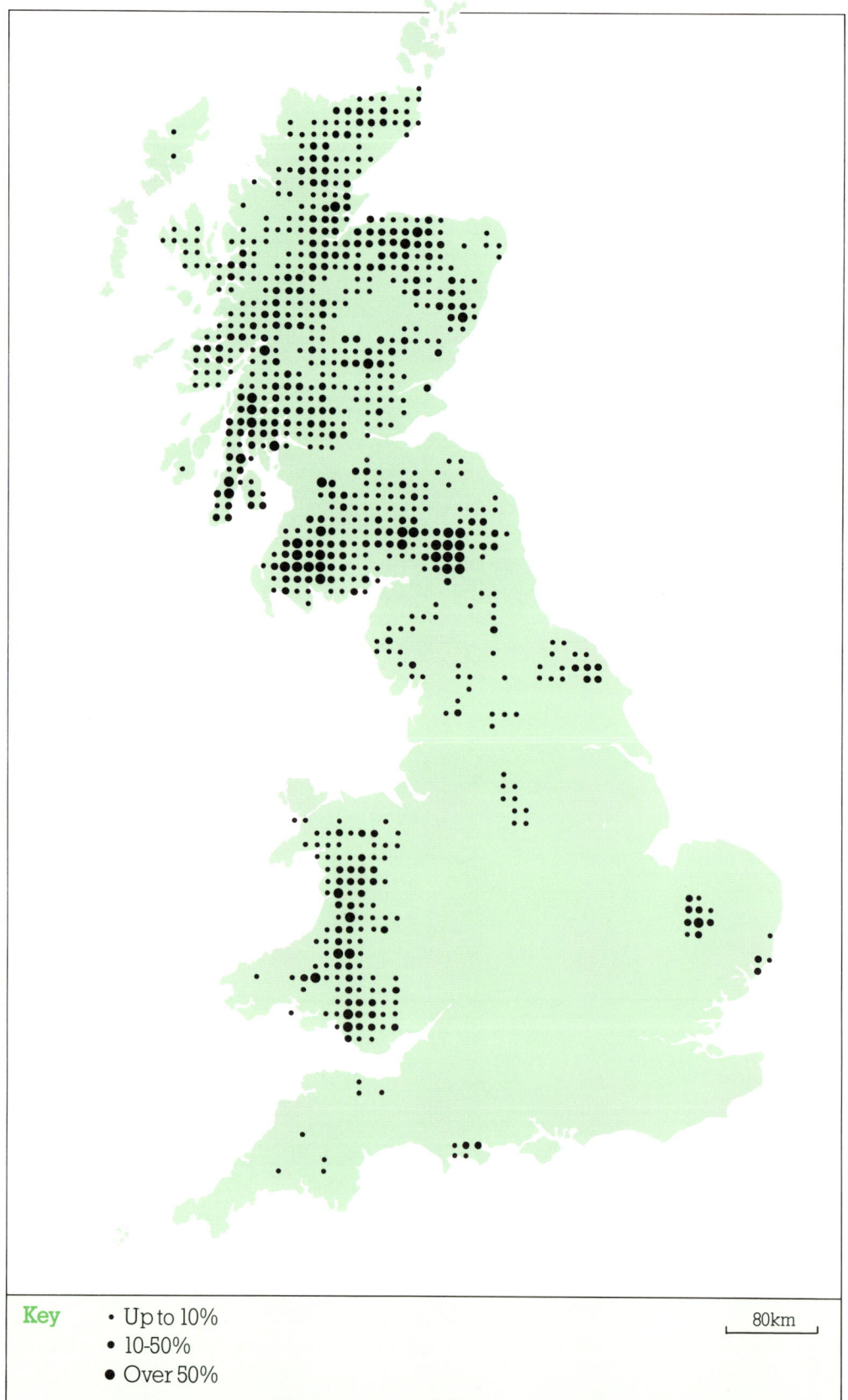

Key	• Up to 10%
	• 10-50%
	● Over 50%

80km

Map 3 The distribution of new conifer forest in Knapdale and Kintyre in 1986. Note: within the total extent of forest in this district (c. 55,000 ha) only about 2,000 ha are semi-natural broadleaved woodland.

Key
Arable land
Afforestation

16km

Maps and tables

Map 4 The world distribution of blanket bog.

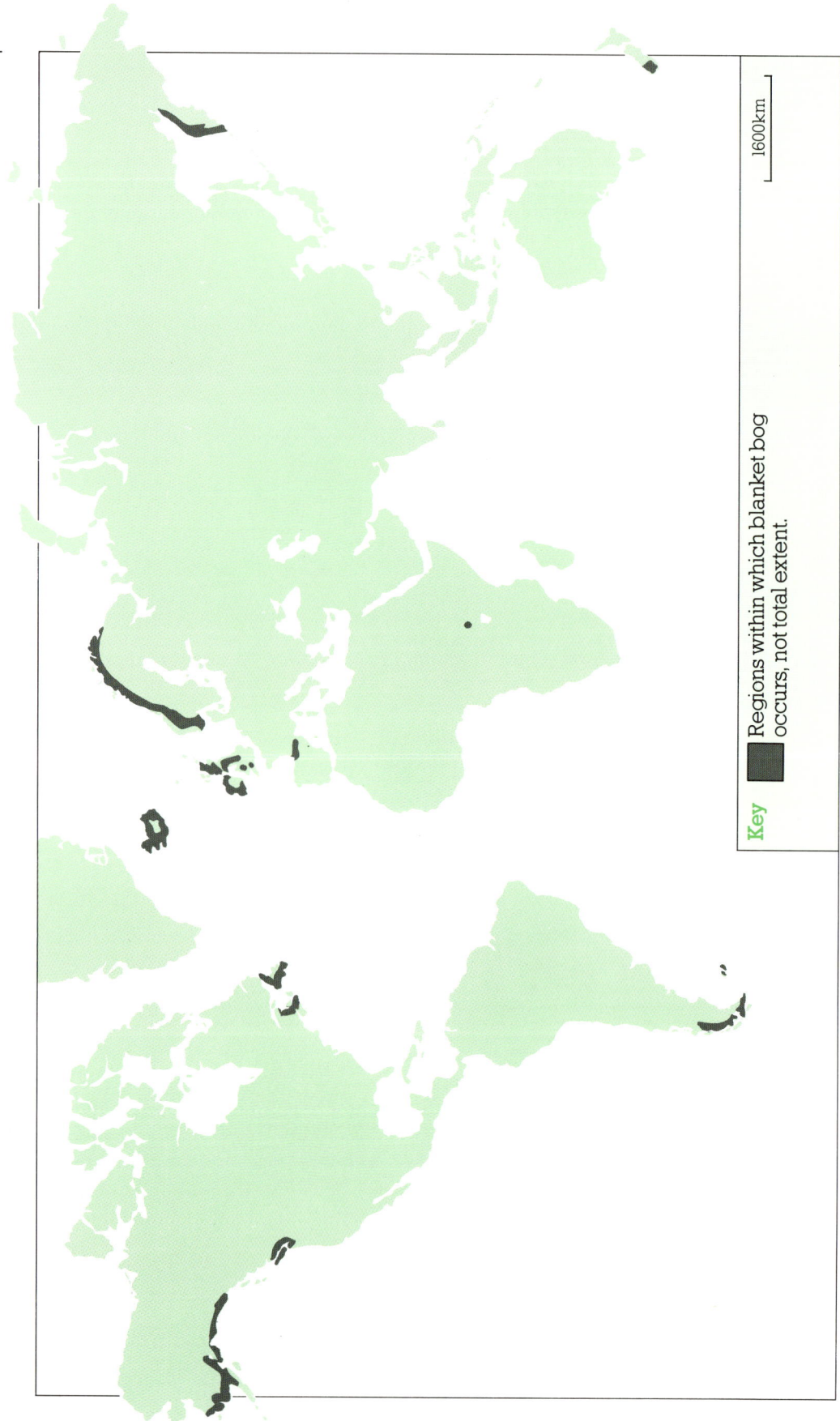

1600km

Key Regions within which blanket bog occurs, not total extent.

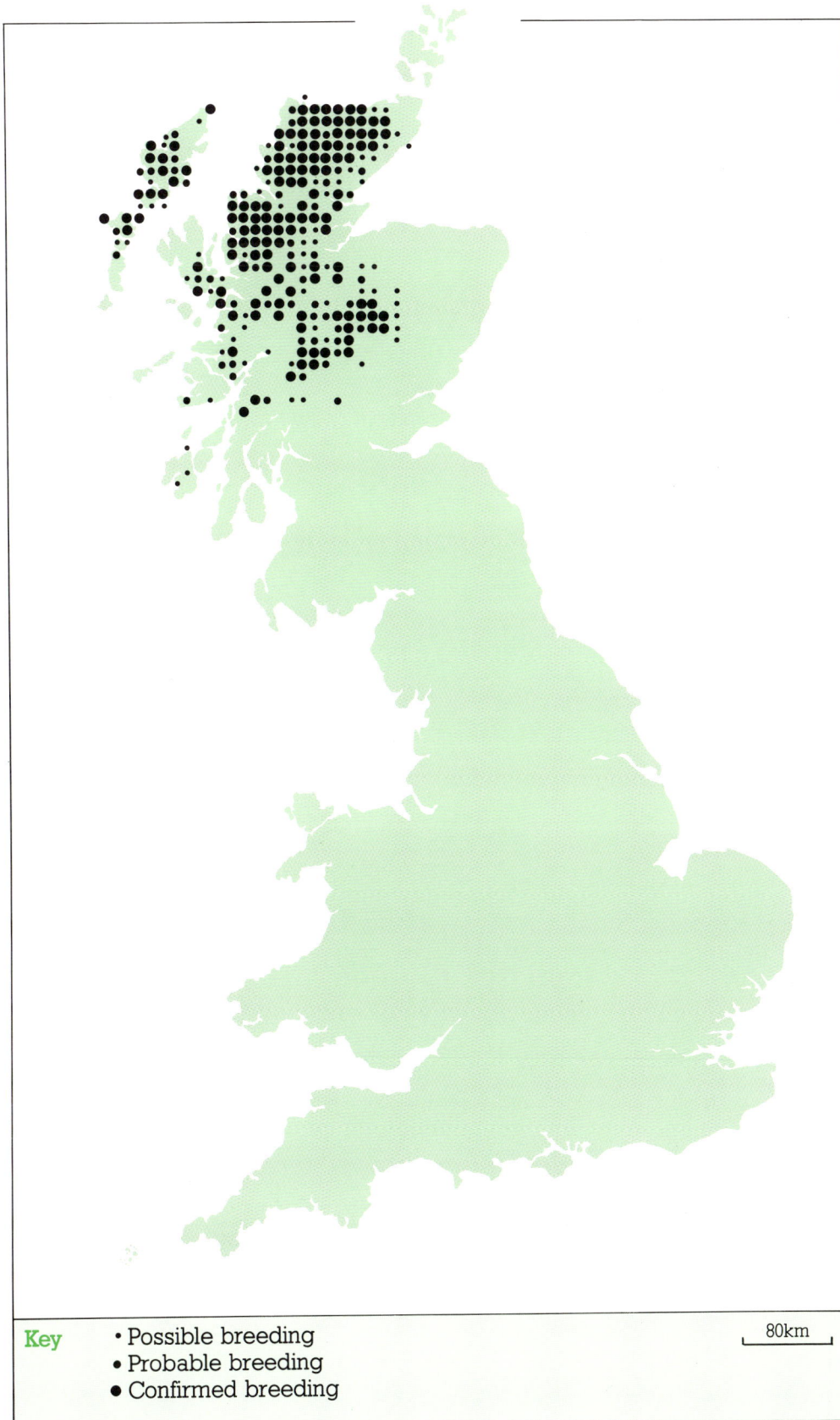

Map 5 The British breeding distribution of the greenshank.

Key
- • Possible breeding
- • Probable breeding
- ● Confirmed breeding

80km

Map 6 The British breeding distribution of the merlin.

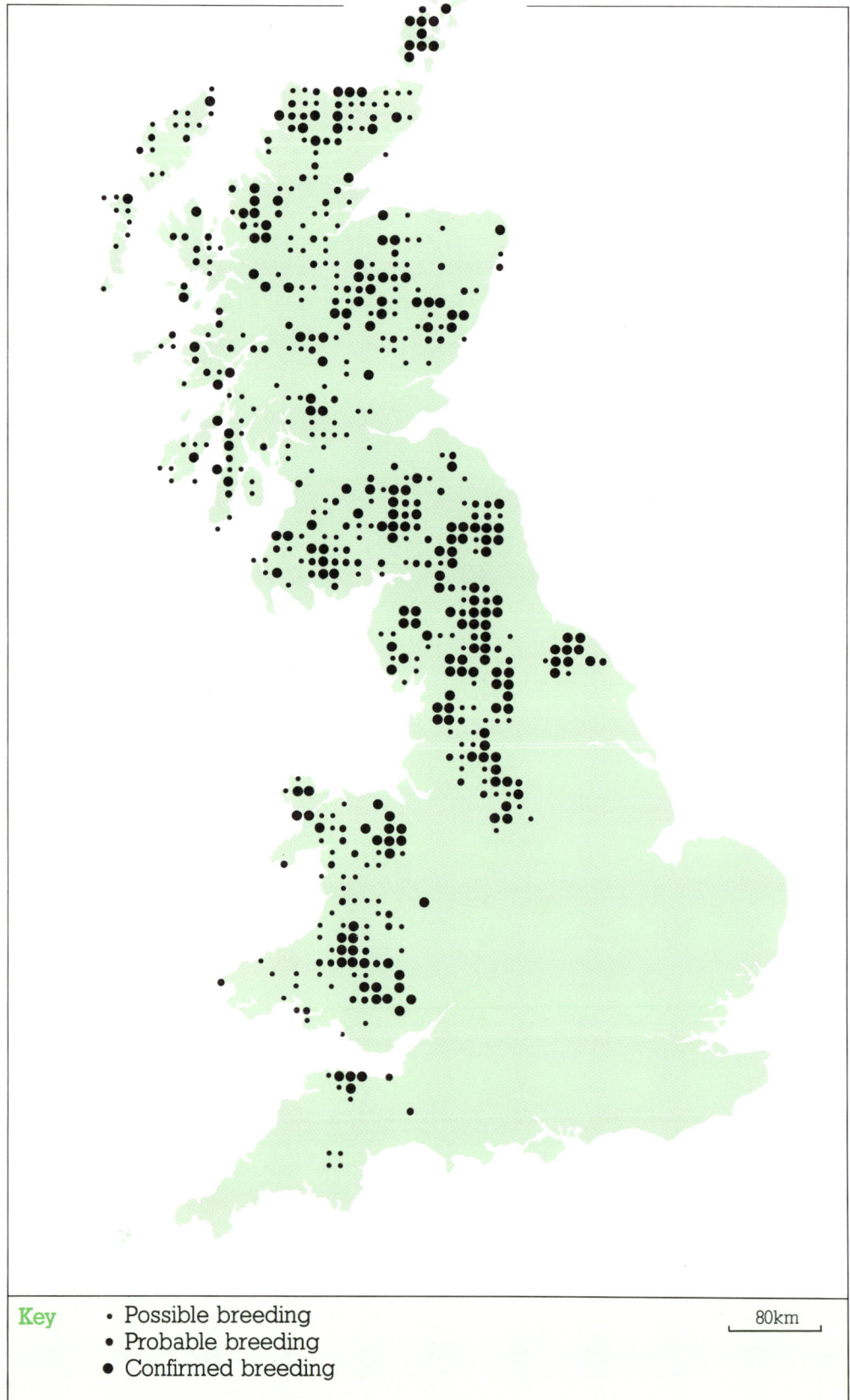

Key
- Possible breeding
- Probable breeding
- Confirmed breeding

80km

Map 7 The European
breeding distribution of
the golden plover.

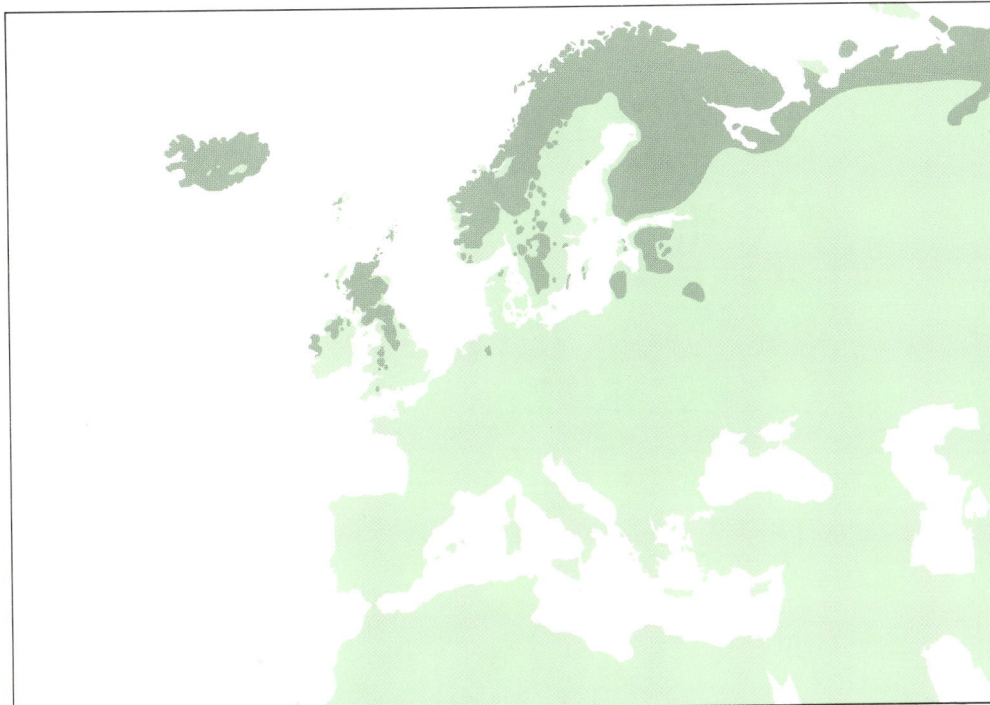

Key — European distribution of golden plover | 800km

Map 8 The world distribution of the Dartford
warbler.

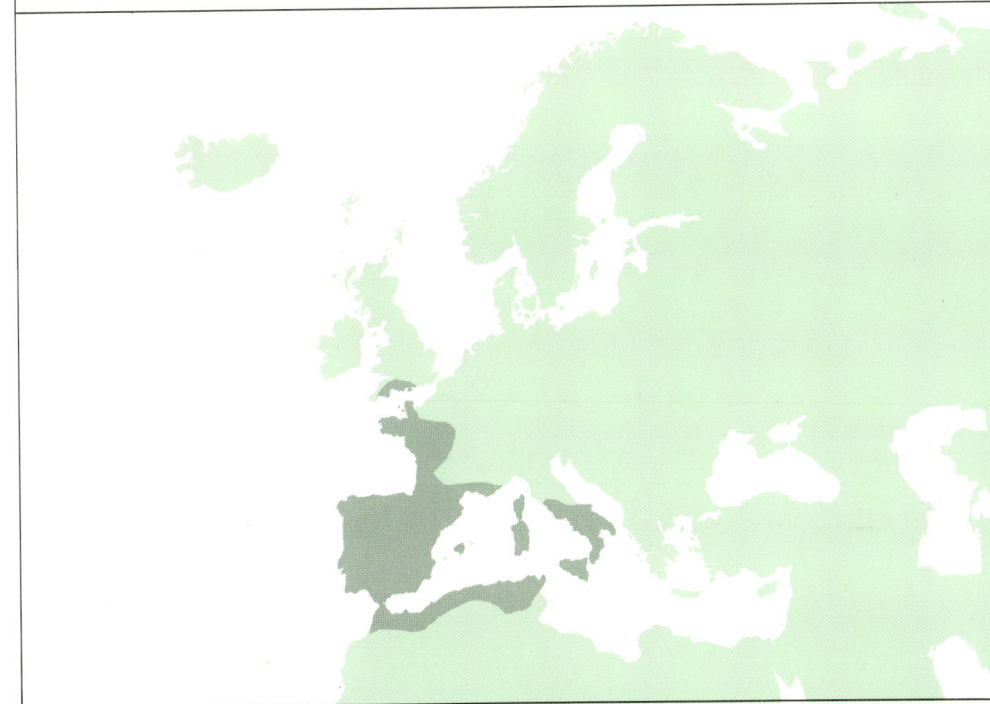

Key — World distribution of Dartford warbler | 800km

Maps and tables

Map 9 The British distribution of the large heath butterfly.

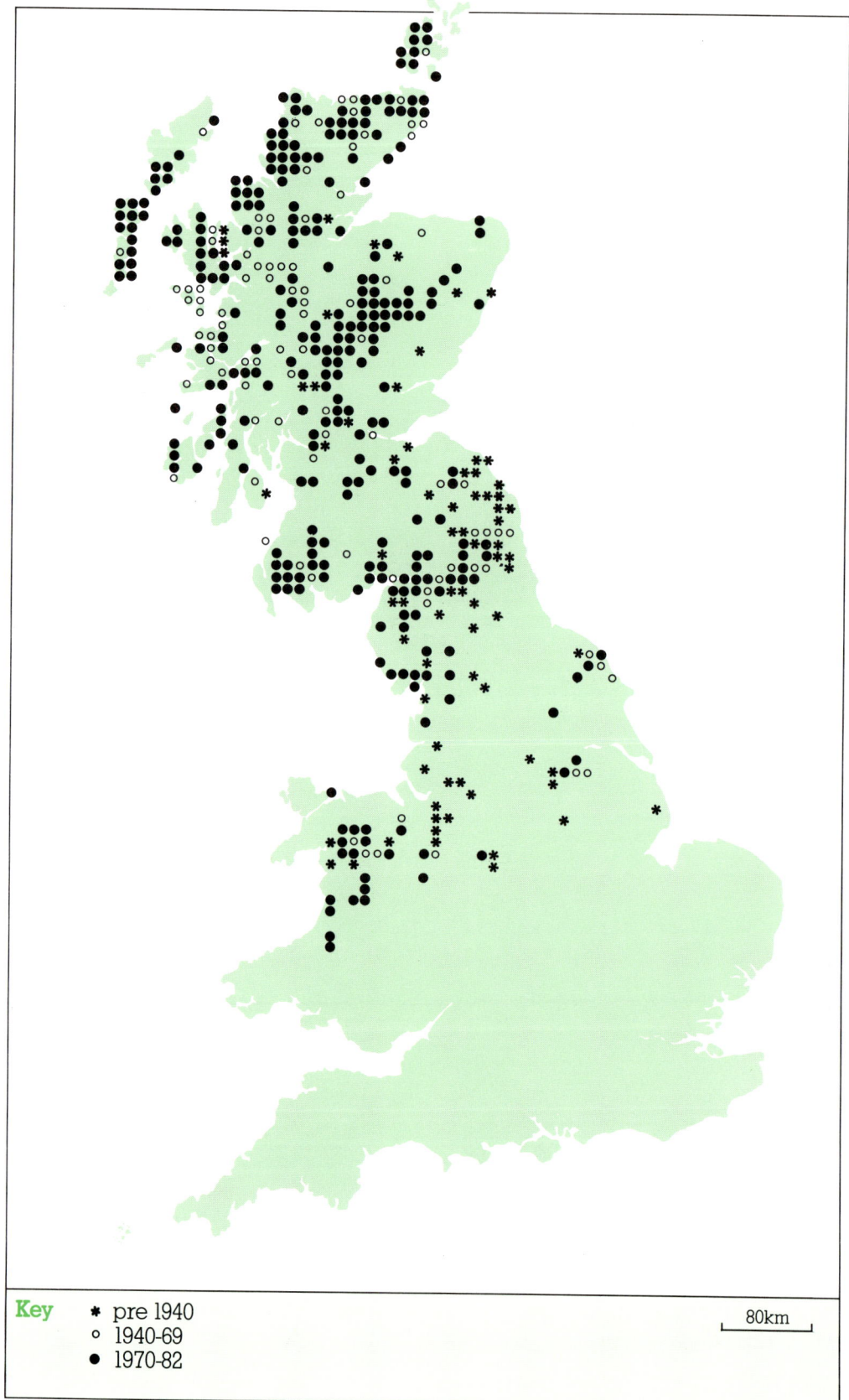

Key
* pre 1940
o 1940-69
● 1970-82

80km

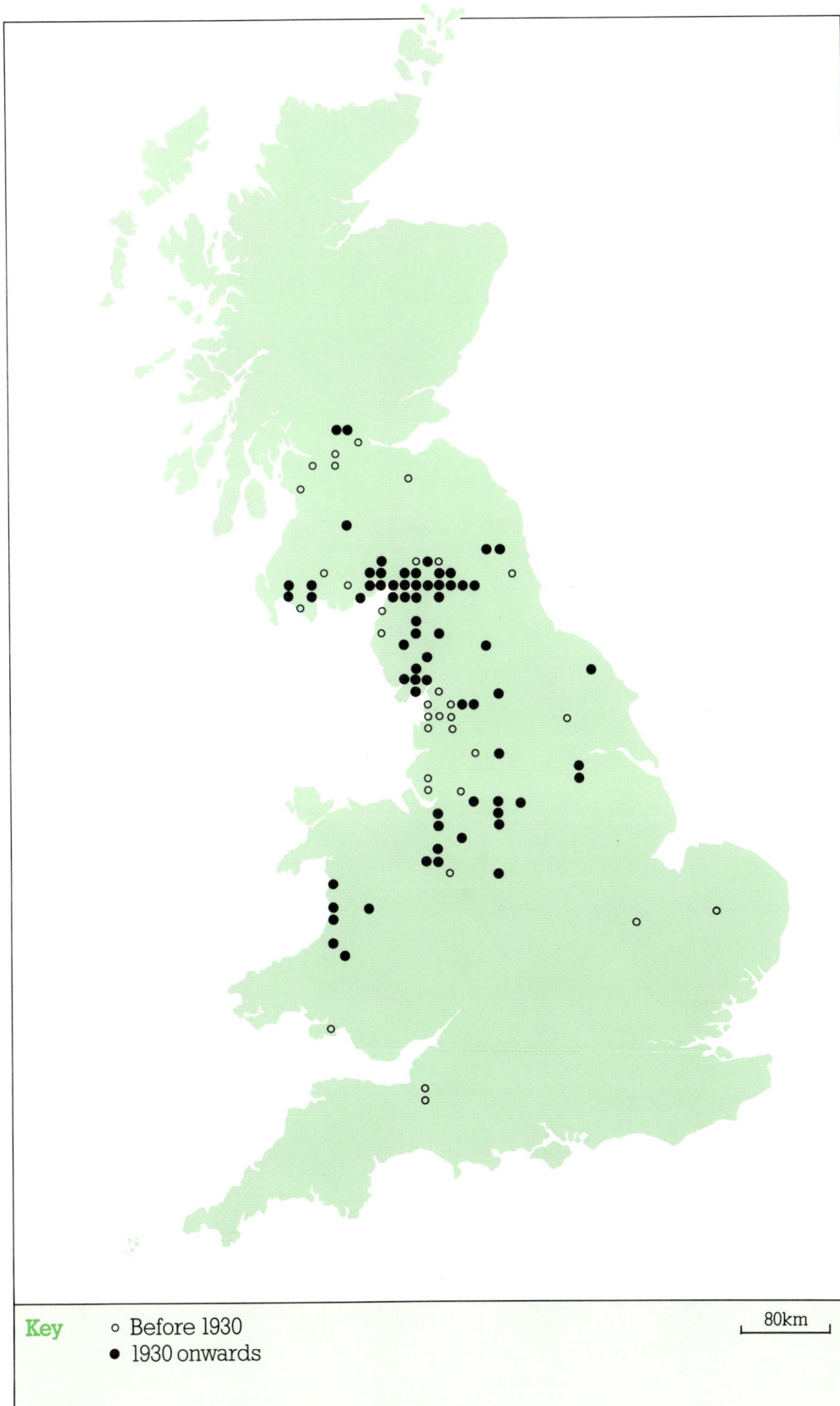

Map 10 The British distribution of bog rosemary *Andromeda polifolia*.

Key ○ Before 1930
     ● 1930 onwards

80km

Maps and tables

Map 11 The British distribution of tall bog-sedge *Carex magellanica*.

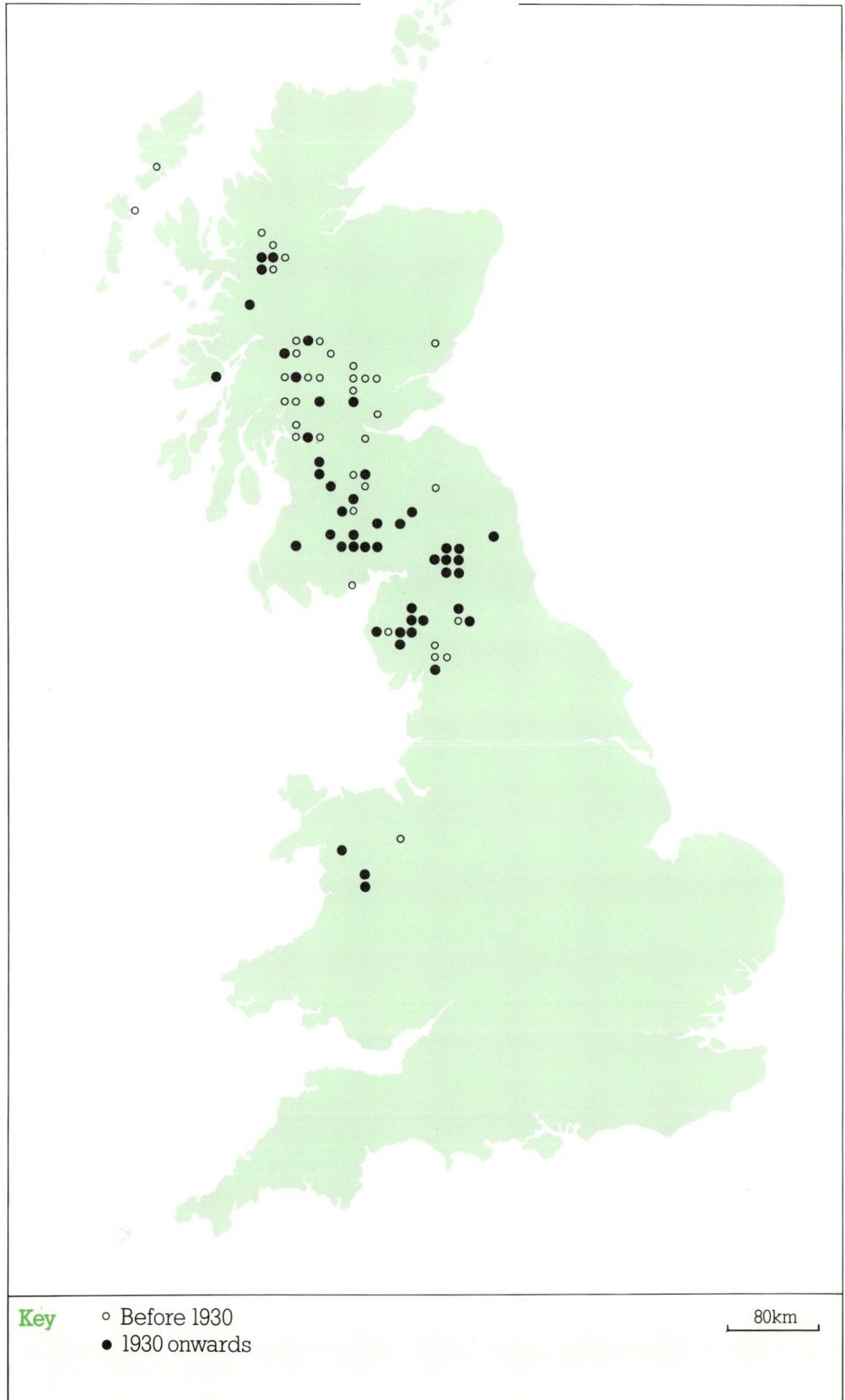

Key

○ Before 1930
● 1930 onwards

80km

200

180   160

220   140

180

160

140

120
120

120

120

160

120

160

120

140

80km

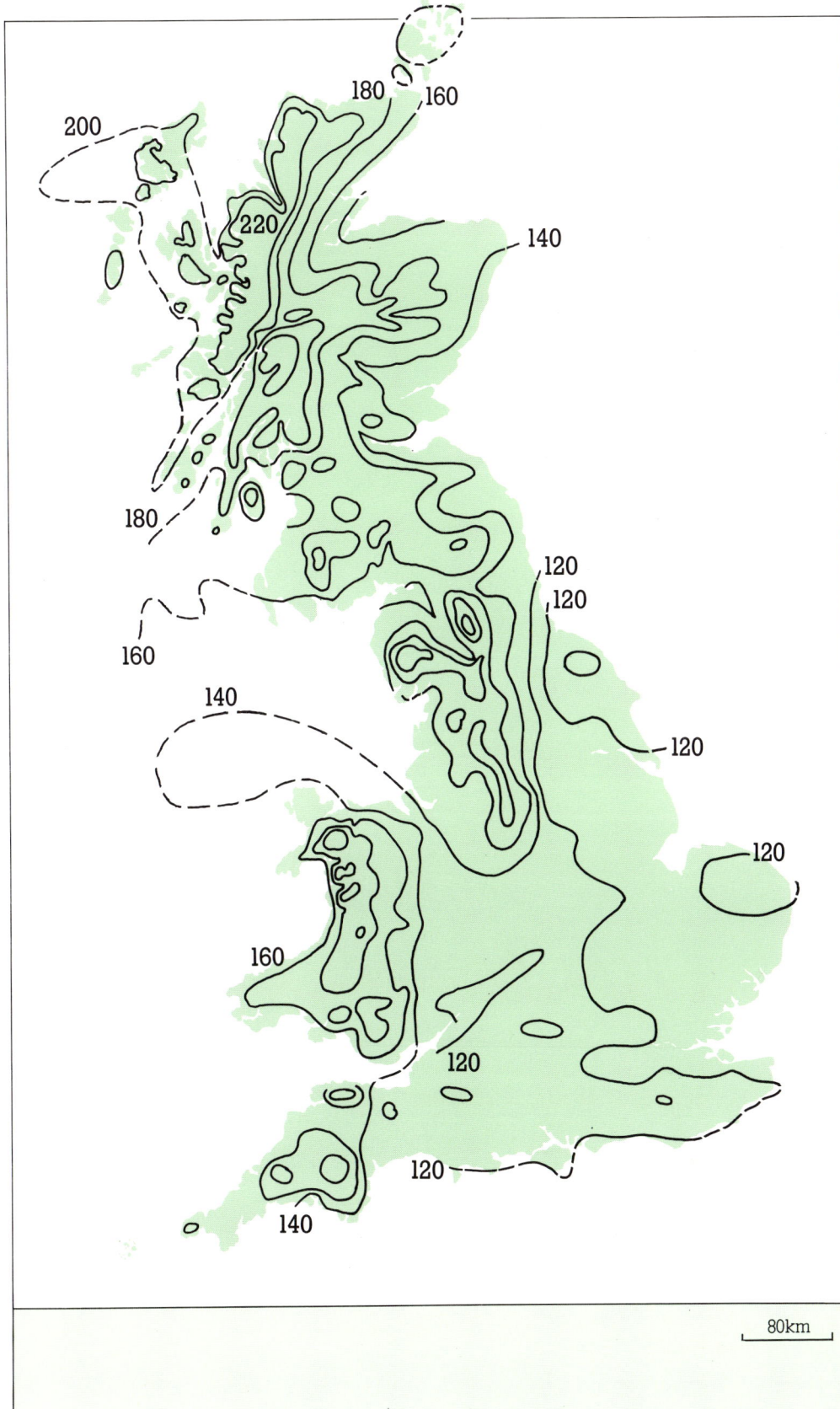

Maps 12, 13 and 14 show
that the patterns of
wetness, temperature and
wind combine to produce
an overall gradient of
decreasing suitability of
climate for tree growth
towards the north-west of
Scotland. The equable
temperature regime of
western areas favours
tree growth in sheltered
places, but on exposed
ground wind becomes
the dominant factor for
trees in the oceanic British
climate. Extreme wetness
and low temperature in
the north and west give an
increasing tendency to
ground waterlogging and
acidic peat formation.

Map 12 The distribution
of "wet days". Note: a wet
day is the meteorological
category of a period of 24
hours within which there
is a precipitation of at
least 1mm. It is a better
index of ecological
wetness of climate than
total rainfall.
Compiled by NCC from
data published in British
Rainfall 1951-60.

Maps and tables

Map 13 Mean temperature (°C) for the warmest month (July). Note: these isotherms are a general measure of summer warmth and show the south to north temperature gradient. Figures are corrected to sea level.

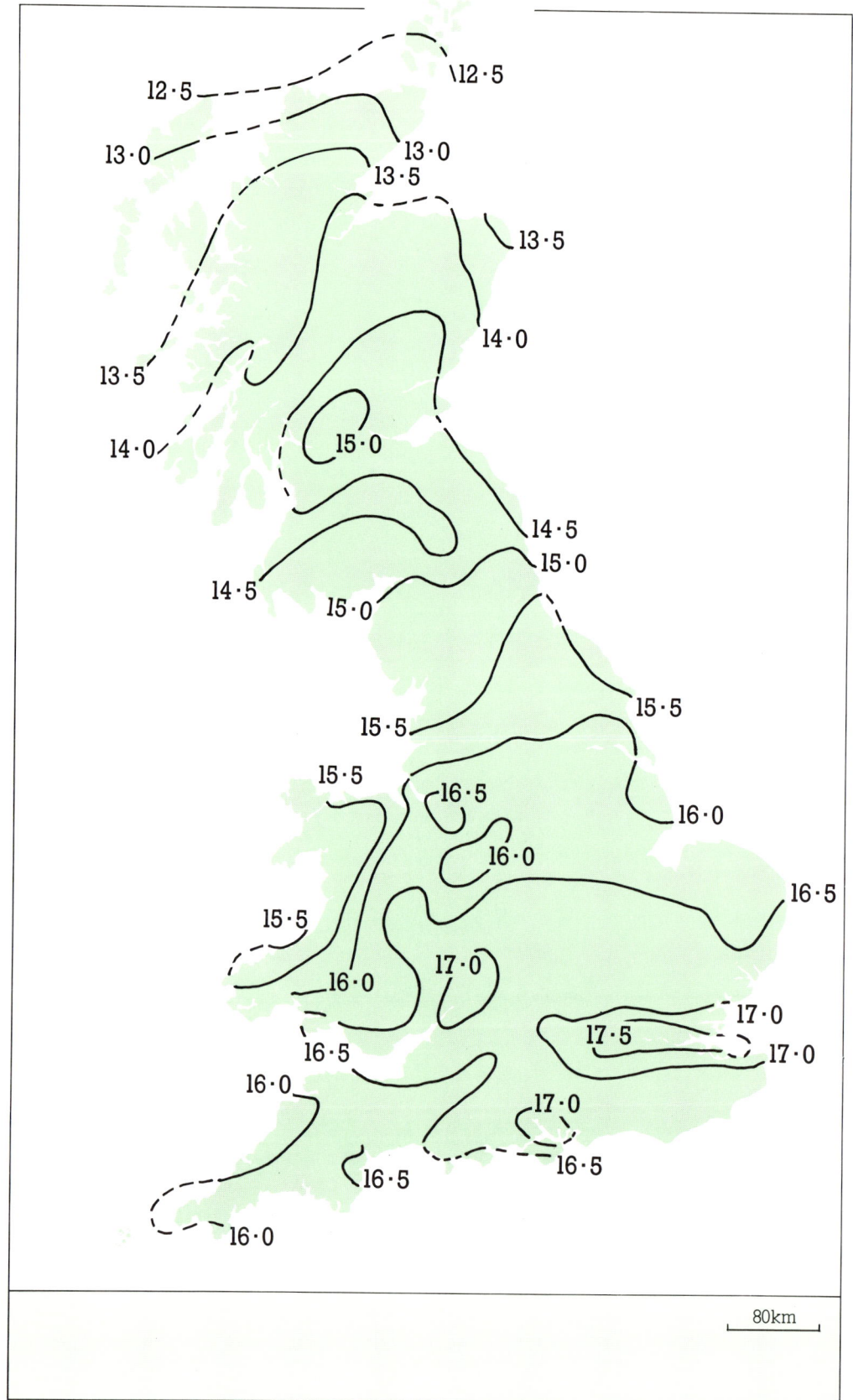

About 12·0

12·5

12·5

13·0

13·0

13·5

13·5

13·5

14·0

13·5

15·0

14·0

14·5

15·0

14·5

15·0

15·5

15·5

15·5

16·5

16·0

15·5

16·0

16·5

15·5

16·0

17·0

17·0

17·5

17·0

16·5

17·0

16·0

16·5

17·0

16·5

16·0

80km

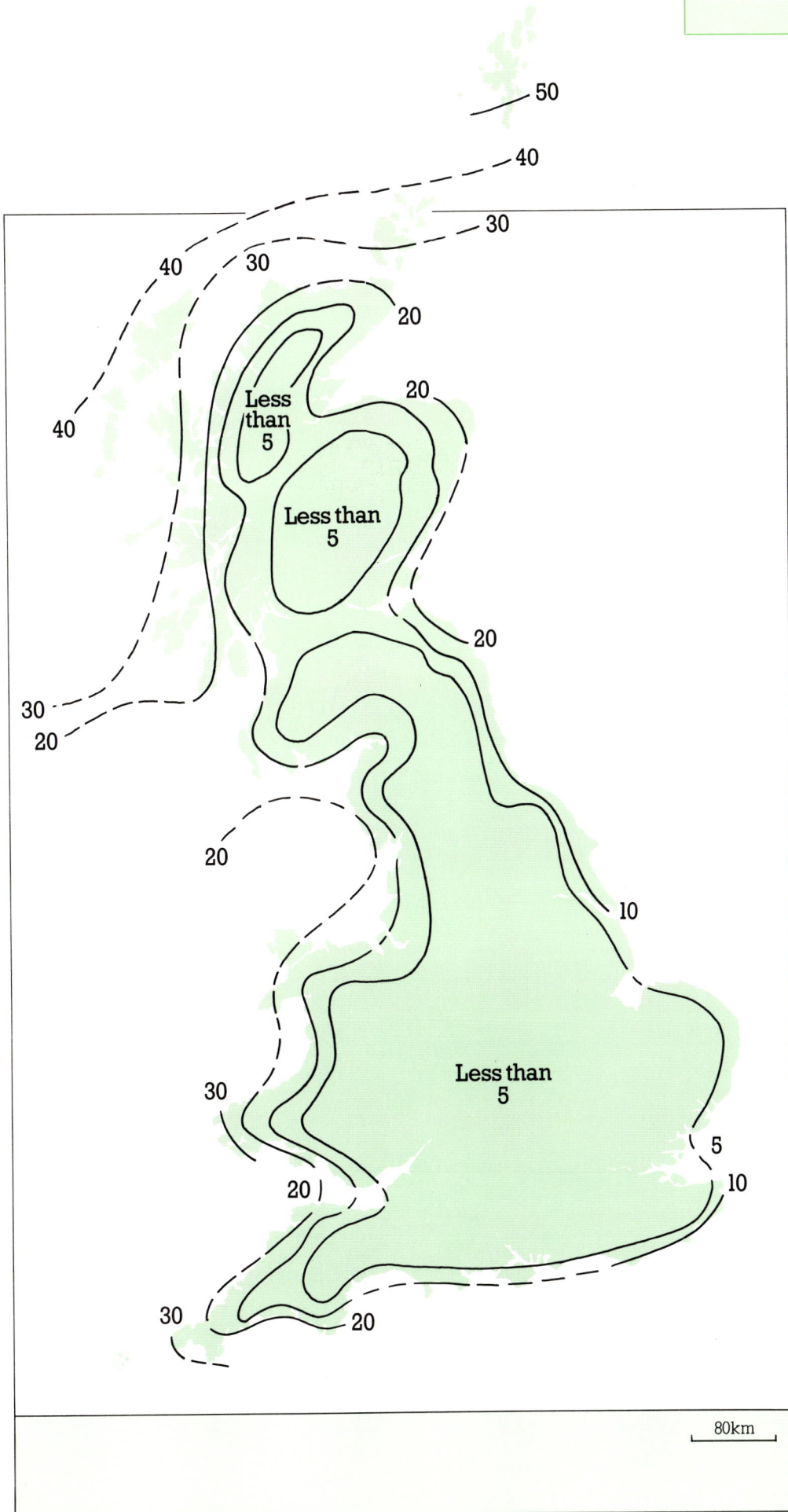

Map 14 Average annual number of days with gale. Note: the isolines show that tendency to high wind increases towards coasts generally, but also towards the north-west of Scotland. Figures are for low altitude recording stations.

50

40

30

40   30

40   20

Less   20
than
5

Less than
5

20

30

20

20

10

Less than
5

30   5

10

20

30   20

80km

Maps and tables

Table 1 Areas of new and afforestable land up to 1985.

	England	Wales	Scotland	Great Britain
			(thousands of hectares)	
Gross area of heath, moorland and hill	1150	700	4400	6250
Area already in forest	207	164	811	1182
Unsuitable for forest (too high)	90 (75)	105 (55)	2000 * (950)	2195 (1080)
Plantable area remaining	853	431	1589	2873
Area of heath, moorland and hill SSSI	172	72	279	523 †

Notes:
Data for the first four categories are based on those in Table 1, Appendix IX of the CAS (1980) study, but revised to take account of further planting and evident relaxation over what is now regarded as plantable land. The area of this land in Scotland (*) now regarded as unsuitable seems too high, and the total area of plantable land still remaining is probably well over 3 Mha. The figures refer to types of land traditionally regarded as afforestable: if afforestation moves downhill onto enclosed farmland, the category and area of plantable land will need substantial revision.

†Approximately sixty per cent of category five is land which might have been regarded as plantable.

Table 2 Effects of afforestation on breeding birds.

Abundance class	Species
Open ground species displaced and permanently reduced	
5	Red-necked phalarope
4	Stone-curlew, Dartford warbler†, golden eagle, merlin, hen harrier*†, chough, wigeon
3	Dunlin, golden plover, ringed plover, greenshank, short-eared owl*†, Arctic skua, raven
2-3	Ring ouzel
2	Curlew*, golden plover, redshank, snipe, wheatear, lesser black-backed gull, common gull, black-headed gull, dipper
1	Lapwing, red grouse*, stock dove
0	Skylark*
Open ground species increasing during the first ten years and/or returning in good numbers in restocked ground	
4	Woodlark
3	Nightjar, black grouse
2	Stonechat, whinchat, grasshopper warbler, twite, cuckoo, mallard
1-2	Tree pipit
1	Grey partridge, pheasant, reed bunting
0	Meadow pipit, yellowhammer
Tree- or cliff-nesting species able to breed within forest but needing open ground for hunting	
5	Red kite
4	Peregrine (merlin sparingly)
3	Buzzard
2	Kestrel
1	Carrion crow, magpie
Wetland species at risk from afforestation	
6	Wood sandpiper
5	Common scoter
4	Greylag goose, black-throated diver
3-4	Goosander
3	Red-throated diver, teal, red-breasted merganser
2	Grey wagtail, common sandpiper
Woodland or scrub species which colonise and increase	
5	Goshawk, firecrest
3	Crossbill, long-eared owl
2	Sparrowhawk, tawny owl, woodcock, siskin, wood warbler, pied flycatcher
1	Long-tailed tit, coal tit, redstart, blackcap, garden warbler, treecreeper, goldcrest, redpoll, bullfinch, greenfinch, whitethroat, chiffchaff, linnet, mistle thrush, jay
0	Woodpigeon, willow warbler, great tit, blue tit, song thrush, blackbird, robin, dunnock, chaffinch, wren

Abundance classes

Number of breeding pairs	
0	>1 million
1	100,000-1,000,000
2	10,000-100,000
3	1,000-10,000
4	100-1,000
5	10-100
6	1-10

* May return in small numbers to restocked ground
† Temporary increase during first 10 years of first rotation

Nature conservation and afforestation

11

Maps and tables

Table 3 Local species of vascular plants which have declined through afforestation.

Huperzia selago	D. intermedia	Crepis paludosa
Lycopodium clavatum	Carum verticillatum	Listera cordata
Lycopodiella inundata	Betula nana	Gymnadenia conopsea
Selaginella selaginoides	Salix repens	Orchis mascula
Equisetum sylvaticum	Arctostaphylos uva-ursi	Dactylorhiza fuchsii
Botrychium lunaria	A. alpinus	D. purpurella
Phegopteris connectilis	Erica ciliaris	Eriophorum latifolium
Dryopteris carthusiana	Vaccinium oxycoccos	Eleocharis multicaulis
Gymnocarpium dryopteris	Pyrola media	Schoenus nigricans
Juniperus communis	Primula farinosa	Rhynchospora alba
Trollius europaeus	Trientalis europaea	Carex hostiana
Viola lutea	Gentiana pneumonanthe	C. lepidocarpa
Hypericum elodes	Menyanthes trifoliata	C. pallescens
Helianthemum nummularium	Cuscuta epithymum	C. limosa
Dianthus deltoides	Pedicularis palustris	C. magellanica
Geranium sylvaticum	Pinguicula lusitanica	C. lasiocarpa
Genista anglica	Utricularia intermedia	C. curta
Ulex gallii	U. minor	C. pauciflora
U. minor	Valeriana dioica	C. dioica
Parnassia palustris	Cirsium helenioides	Festuca vivipara
Drosera anglica	Antennaria dioica	Agrostis curtisii

	Plant ▼	Thicket forest ⟶ Maturing forest	Fell & re-stock ▼	
Unplanted ground	Young, pre-thicket forest		Young, pre-thicket forest	Thicket forest
Ecosystem				
Dwarf shrub heath Grassland Peat bog Sand dune	Dense planting mainly of Sitka spruce, lodge-pole pine and larch Remaining open ground fragmentary, but variable according to topography	Development and duration varies according to conditions (especially climate) and tree species, which affect silvicultural regime Thinning least, thicket stage longest and rotation shortest in north and west, and at high altitude: opposite trends to south and east and at low altitude	Return to situation of first rotation, but opportunity for diversification of age class pattern and tree species	
Land use				
Grazing range Sheep Cattle Ponies Feral goats Red deer Red grouse	Stock removed Deer and goats controlled Burning ceases Ground fertilising Ploughing usual Herbicides used locally	Fertilisers sometimes applied Insecticides used locally Predator control varies: protected species benefit Brashing and thinning of trees variable, as above Windthrow locally causes premature clearance	Ground preparation and forest establishment phase Deer foraging may give problems	Much the same as first rotation and again variable, depending on how much of the opportunities for ecological diversification are exploited.
Vegetation				
Ericaceous shrubs Grass, sedge and rush Bog moss, deer sedge, cotton grass Dune heath	Vegetation more luxuriant Dwarf shrubs increase, wet ground plants decline	Field layer decreases rapidly to virtual extinction in dense thickets: re-development depends on thinning and light penetration. Often only a species-poor version of semi-natural woodland community	Some recovery of previous dwarf shrubs and grassland but variable. Few peatland species and new invaders appear Slash often thick	
Animals				
Moorland birds, especially waders and predators Red deer, hares, rabbits Insect fauna rich	Waders decline Voles and their predators, and woodland edge species increase	Bird community dominated by woodland songbirds develops. Presence/abundance of larger species depends on thinning Some mammals increase, including pine marten and wildcat locally, and both red deer and roe deer need controlling Local decline of salmonid fish	Mainly a woodland edge bird community, few waders, feeding value for deer improves Some insects benefit	
Abiotic conditions				
Mainly infertile soils; acidic brown earths, podsols, gleys and blanket peats Stable hydrology	Water run-off and soil erosion usually much increased Fertilisers may increase nutrient levels of soils and run-off water	Water run-off becomes reduced below original level by interception effect. Soil erosion declines, acidic humus layer increases and podsolisation is enhanced. Scavenging of atmospheric pollutants increases acidity of water in rivers and lakes	Tends to return to previous state of first rotation	

Notes:
For most of the new forests established on open ground, we cannot yet observe effects beyond the early stages of the second rotation, and the bulk of the area is still in the first rotation.

Table 4 The ecological succession produced by afforestation.

107

Acknowledgments

Acknowledgements for the use of information for maps and diagrams to:

Penguin Books Ltd.
Biological Records Centre
Oxford University Press
T. & A. D. Poyser Ltd.
Elsevier Nederland B.V.
Meteorological Office

Photos supplied by NCC, Ardea,
E. Hosking, Frank Lane and
Nature Photographers.